Modeling and Optimization in Science and Technologies

Volume 21

AF183758

The book series Modeling and Optimization in Science and Technologies (MOST) publishes basic principles as well as novel theories and methods in the fast-evolving field of modeling and optimization. Topics of interest include, but are not limited to: methods for analysis, design and control of complex systems, networks and machines; methods for analysis, visualization and management of large data sets; use of supercomputers for modeling complex systems; digital signal processing; molecular modeling; and tools and software solutions for different scientific and technological purposes. Special emphasis is given to publications discussing novel theories and practical solutions that, by overcoming the limitations of traditional methods, may successfully address modern scientific challenges, thus promoting scientific and technological progress. The series publishes monographs, contributed volumes and conference proceedings, as well as advanced textbooks. The main targets of the series are graduate students, researchers and professionals working at the forefront of their fields.

Indexed by SCOPUS. All books published in the series are submitted for consideration in Web of Science.

J. Juan Rosales García ·
Jesús Enrique Escalante-Martínez ·
Francisco Antonio Godínez Rojano ·
Jessica Carmin Mendiola Fuentes · Lizeth Torres
Editors

Advances in Fractional Calculus

Springer

Editors
J. Juan Rosales García
Departamento de Ingeniería Eléctrica
Universidad de Guanajuato
Salamanca, Guanajuato, Mexico

Jesús Enrique Escalante-Martínez
Facultad de Ingeniería Mecánica y Eléctrica
Universidad Veracruzana
Poza Rica, Veracruz, Mexico

Francisco Antonio Godínez Rojano
Instituto de Ingeniería
Universidad Nacional Autónoma de México
Mexico City, Mexico

Jessica Carmin Mendiola Fuentes
Universidad del Caribe
Cancún, Mexico

Lizeth Torres
Instituto de Ingeniería
Universidad Nacional Autónoma de México
Mexico City, Mexico

ISSN 2196-7326 ISSN 2196-7334 (electronic)
Modeling and Optimization in Science and Technologies
ISBN 978-3-031-83991-7 ISBN 978-3-031-83989-4 (eBook)
https://doi.org/10.1007/978-3-031-83989-4

This Springer imprint is published by the registered company Springer Nature Switzerland AG
The registered company address is: Gewerbestrasse 11, 6330 Cham, Switzerland

Preface

It is an honor to present this volume that compiles the outstanding presentations of the 6th Mexican Workshop on Fractional Calculus 2024, which took place at the sciences and engineering division of the University of Guanajuato, León, Guanajuato, México. The international Mexican Workshops on Fractional Calculus (MWFC) were born in the Engineering Division of the Irapuato-Salamanca Campus of the University of Guanajuato in 2012. Their main objective is to present the latest theoretical advances and applications of fractional calculus to different areas of Science and Engineering. These conferences have been a meeting point for researchers, academics and graduate students interested in the field of fractional calculus, promoting the exchange of ideas and advances in this fascinating area.

The selection of the 13 papers that make up this book was carried out with a rigorous approach, considering both the quality and the length of the manuscripts. To ensure an efficient and transparent review process, we used the EasyChair platform, which facilitated the organization and management of the evaluation process. The reviewers were chosen based on their expertise in fractional calculus, ensuring that each had not only the necessary technical knowledge, but also a deep understanding of the specific topic of each article. This allowed the evaluation to be carried out by true experts in the field. The review of each chapter was carried out through a peer review process, a fundamental standard in scientific publishing that guarantees the quality and validity of the papers presented.

We are convinced that this volume not only contributes to the advancement of fractional calculus but will also serve as a valuable source of information and reference for researchers and students interested in this constantly evolving field. We thank all the authors, reviewers and members of the organizing committee for their dedication and effort in making this book possible.

Enjoy the reading!

The Editorial Committee
Salamanca, Mexico — J. Juan Rosales García
Poza Rica, Mexico — Jesús Enrique Escalante-Martínez
Mexico City, Mexico — Francisco Antonio Godínez Rojano
Cancún, Mexico — Jessica Carmin Mendiola Fuentes
Mexico City, Mexico — Lizeth Torres

Contents

Chapter 1
A Fundamental Theorem of Curves in the Euclidean Space with Caputo Fractional Derivatives

Roberto Carlos Balcázar-Araiza and José Matías Navarro-Soza

1.1 Introduction

1.1.1 Background

The *Fractional Calculus*, loosely speaking, comprehends the study, analysis and application of *fractional differential and integral operators*, where the foregoing ones result from the extension of iterated derivatives and integrals of integer order to real or complex orders.

On the other hand, in *Differential Geometry*, it is important to know which is the minimal amount of information we need to characterize a curve in the n-dimensional Euclidean space; the answer to this query is given by the *Fundamental Theorem of Curves in* $\mathbb{R}^n$, which states that given a finite number of curvature functions and initial conditions, there is a unique parametric curve (up to rigid motions) whose curvatures are the given functions. It is believed that L. Euler gave a first proof of this result in [1] for plane curves; a century later, R. Hoppe proved the theorem in [2] for spatial curves after the works of Frenet and Serret ([3, 4], respectively).

The interaction between Fractional Calculus and Differential Geometry has been studied within the last 15–20 years, approximately. With regard to surfaces, an approach to study viscoelastic models through Riemann-Liouville and Caputo fractional derivatives was conducted in [5, 6], where generalizations of the fundamental forms, the Gaussian and mean curvatures, Christoffel symbols, connections and the Riemannian and torsion tensors were proposed. In relation to curves, more work has been done with the goal of generalizing the notions of the arc length, the Frenet

R. C. Balcázar-Araiza (✉) · J. M. Navarro-Soza
Universidad Autónoma de Yucatán, Mérida, Yucatán, México
e-mail: a20216411@alumnos.uady.mx

J. M. Navarro-Soza
e-mail: matias.navarro@correo.uady.mx

J. J. Rosales García et al. (eds.), *Advances in Fractional Calculus*,
Modeling and Optimization in Science and Technologies 21,
https://doi.org/10.1007/978-3-031-83989-4_1

curvatures and the Frenet-Serret formulae; for example, Lazopoulos and Lazopoulos proposed a study in [7] of the fractional differential geometry of curves by means of the *L-fractional derivative*, which is a quotient of Caputo derivatives; however, their treatment of the fractional chain rule is unsatisfactory, in the same manner as the alternative research from Yajima et al. in [8], where the study of the geometry of curves in $\mathbb{R}^3$ is conducted with the classical Caputo derivative, but many conveniences are obtained by forcing the fractional chain rule for Caputo derivatives to adopt a similar form to the classical chain rule without justification; this problem is inherited by the works done by Aydın et al. and Öğrenmiş in [9, 10], respectively. On the other hand, Lopez and Rubio proposed in [11] to study planar curves throughout the decomposition of the vector of fractional derivatives, in terms of the classical Frenet-Serret frame.

From the geometrical point of view and the perspectives drawn for plane curves in [11], it is clear for us that the introduction of non-local fractional derivatives for the study of the intrinsic and extrinsic geometry of curves, or manifolds in general, must carry on the loss of desirable properties, such as the invariance under reparametrizations; although this may seem like a flaw for the theory, we believe that this is a natural consequence from the fact that vectors related to fractional derivatives will not necessarily live in the tangent bundle to a manifold anymore (see [11] or [6] for a better understanding of this conjecture). Therefore, we consider that a theory of fractional differential geometry must adapt to the rules of the considered non-local fractional operators; with that spirit in mind, and mainly inspired by the work done in [5, 6, 11], we provide another way to work with the classical Caputo differential operator which does not rely on the chain or product rule, by means of combining the use of fractional and ordinary derivatives.

For the rest of this section, we have gathered the basic tools from Caputo fractional calculus that we will use. In Sect. 1.2, we extend basic notions from the classical theory of curves when the Caputo fractional derivative is involved and in Sect. 1.3, a Frenet-Serret-type theorem is proved. Finally, we close in Sect. 1.4 with a proof of a fundamental theorem for the local theory of curves based on the Caputo differential operators.

1.1.2 Basics of Fractional Calculus

Definition 1.1 Let $f \in L^1([a,b])$ and $\alpha > 0$. The **Riemann-Liouville left fractional integral of f of order** α is defined as

$$I^{\alpha}_{a+} f(x) := \frac{1}{\Gamma(\alpha)} \int_a^x (x-t)^{\alpha-1} f(t)dt, \tag{1.1}$$

while the **Riemann-Liouville right fractional derivative of f of order** α is given by

$$I_{b-}^{\alpha} f(x) := \frac{1}{\Gamma(\alpha)} \int_x^b (t-x)^{\alpha-1} f(t)dt. \tag{1.2}$$

Besides, if $\alpha = 0$, we adopt the convention $I_{a+}^0 f(x) = I_{b-}^0 f(x) = f(x)$ (based on Theorem 2.7 from [12]).

A main and frequently used property for the fractional integral is known as the *semigroup* property, which is proved in Sect. 2.3 from [12].

Theorem 1.1 *Let $f \in C([a,b])$ and $\alpha, \beta \geq 0$. Then, for every $x \in [a,b]$,*

$$I_{a+}^{\alpha+\beta} f(x) = I_{a+}^{\alpha} I_{a+}^{\beta} f(x) \text{ and}$$
$$I_{b-}^{\alpha+\beta} f(x) = I_{b-}^{\alpha} I_{b-}^{\beta} f(x).$$

Example 1.1 The easiest example for the calculation of a fractional integral corresponds to the case of power functions. Let $f(x) = (x-a)^{\beta}$ with $\beta > -1$. It is proven in Example 2.1 from [13] that if $\alpha > 0$,

$$I_{a+}^{\alpha} f(x) = \frac{\Gamma(\beta+1)}{\Gamma(\beta+\alpha+1)}(x-a)^{\beta+\alpha}. \tag{1.3}$$

In the following statement, a useful relation between the left and right fractional integrals is given if we define the *reflection operator* $Q : L^1([a,b]) \to L^1([a,b])$ by $Qf(x) = f(a+b-x)$.

Proposition 1.1 *For $f \in L^1([a,b])$, it is true that*

$$I_{b-}^{\alpha} f(x) = Q I_{a+}^{\alpha} Q f(x). \tag{1.4}$$

Proof First, we see that

$$\begin{aligned} I_{a+}^{\alpha} Qf(x) &= \frac{1}{\Gamma(\alpha)} \int_a^x (x-t)^{\alpha-1} Qf(t)dt \\ &= \frac{1}{\Gamma(\alpha)} \int_a^x (x-t)^{\alpha-1} f(a+b-t)dt. \end{aligned}$$

If we let $u = a+b-t$, this change of variable transforms the integral into

$$\begin{aligned} \frac{1}{\Gamma(\alpha)} \int_a^x (x-t)^{\alpha-1} f(a+b-t)dt &= \frac{1}{\Gamma(\alpha)} \int_{a+b-x}^b (x-a-b+u)^{\alpha-1} f(u)du \\ &= I_{b-}^{\alpha} f(a+b-x) \\ &= Q I_{b-}^{\alpha} f(x) \end{aligned}$$

and because $Q^2 f(x) = f(x)$, it follows that $I_{b-}^{\alpha} f(x) = Q I_{a+}^{\alpha} Q f(x)$. □

Example 1.2 From Example 1.1 and Proposition 1.1, it can be deduced that for $f(x) = (b-x)^\beta$ with $\beta > -1$,

$$I_{b-}^\alpha f(x) = \frac{\Gamma(\beta+1)}{\Gamma(\beta+\alpha+1)}(b-x)^{\beta+\alpha}. \tag{1.5}$$

A result given in [13], Theorem 2.7, establishes conditions for the limit and fractional integration operations to be interchanged as in the classical case ($\alpha = 1$), which turns out to be useful in the computation of some fractional integrals and derivatives.

Theorem 1.2 *Let $\alpha > 0$ and assume that $(f_k)_{k=1}^\infty$ is a uniformly convergent sequence of continuous functions on $[a,b]$. Then we may interchange the fractional integral operator and the limit process, i.e.*

$$I_{a+}^\alpha\left(\lim_{k\to\infty} f_k\right)(x) = \lim_{k\to\infty} I_{a+}^\alpha f_k(x).$$

In particular, the sequence $(I_{a+}^\alpha f_k(x))$ is uniformly convergent.

Corollary 1.1 *Under the same hypotheses of Theorem 1.2,*

$$I_{b-}^\alpha\left(\lim_{k\to\infty} f_k\right)(x) = \lim_{k\to\infty} I_{b-}^\alpha f_k(x).$$

and the sequence $(I_{b-}^\alpha f_k(x))$ is uniformly convergent too.

Proof It follows from applying Proposition 1.1 to Theorem 1.2 and the simple fact that the reflection operator Q preserves continuity. □

As a consequence of Theorem 1.2, Corollary 1.1 and Examples 1.1 and 1.2,we have the following

Corollary 1.2 *If f has a power series representation within the convergence interval $(a-R, a+R)$ for some convergence radius $R \in (0,\infty]$ given by $f(x) = \sum_{k=0}^\infty \frac{f^{(k)}(a)}{k!}(x-a)^k$, then for any $x \in [a, a+R)$,*

$$I_{a+}^\alpha f(x) = \sum_{k=0}^\infty \frac{f^{(k)}(a)}{\Gamma(k+\alpha+1)}(x-a)^{k+\alpha}.$$

Analogously, if $f(x) = \sum_{k=0}^\infty \frac{f^{(k)}(b)}{k!}(x-b)^k$ within $(b-R, b+R)$, for any $x \in (b-R, b]$ we have

$$I_{b-}^\alpha f(x) = \sum_{k=0}^\infty \frac{(-1)^k f^{(k)}(b)}{\Gamma(k+\alpha+1)}(b-x)^{k+\alpha}.$$

Example 1.3 Let us recall that

$$\sin x = \sum_{k=0}^{\infty} \frac{(-1)^k}{(2k+1)!} x^{2k+1} = x E_{2,2}(-x^2),$$

$$\cos x = \sum_{k=0}^{\infty} \frac{(-1)^k}{(2k)!} x^{2k} = E_{2,1}(-x^2),$$

where

$$E_{r,s}(x) := \sum_{k=0}^{\infty} \frac{x^k}{\Gamma(rk+s)}$$

is the so-called *Mittag-Leffler function on two parameters* $r, s \in \mathbb{R}$. It is known that a sufficient condition for the convergence of $E_{r,s}(x)$ on the whole real line is that $r, s > 0$ (c.f. Theorem 4.1, p. 68 from [13]). By Corollary 1.2, it follows that for any $\alpha > 0$ and $x \geq 0$,

$$I^{\alpha}_{0+}(\sin x) = x^{1+\alpha} E_{2,2+\alpha}(-x^2)$$

and

$$I^{\alpha}_{0+}(\cos x) = x^{\alpha} E_{2,1+\alpha}(-x^2).$$

Some left fractional integrals of the sine and cosine functions can be seen in Fig. 1.1.

For a *right* example, given that $\cos(x) = \cos(2\pi - x)$ and $\sin(x) = -\sin(2\pi - x)$, we can use Corollary 1.2 to deduce that for any $x \leq 2\pi$ and $\alpha \geq 0$,

$$I^{\alpha}_{2\pi-}(\sin x) = -(2\pi - x)^{1+\alpha} E_{2,2+\alpha}(-(2\pi - x)^2)$$

and

$$I^{\alpha}_{2\pi-}(\cos x) = (2\pi - x)^{\alpha} E_{2,1+\alpha}(-(2\pi - x)^2).$$

In Fig. 1.2, we show some particular cases.

From Definition 1.1, a widely known notion of the fractional derivative can be deduced.

Definition 1.2 For $f : [a, b] \to \mathbb{R}$, let $\alpha \geq 0$ and $n \in \mathbb{N}$ such that $n - 1 < \alpha \leq n$. The **Caputo left fractional derivative of order** α is defined by

$$D^{\alpha}_{a+} f(x) := I^{n-\alpha}_{a+} f^{(n)}(x), \tag{1.6}$$

while the **Caputo right fractional derivative of** f **of order** α is expressed as

$$D^{\alpha}_{b-} f(x) := (-1)^n I^{n-\alpha}_{b-} f^{(n)}(x). \tag{1.7}$$

Corollary 1.3 *The following relationship between left and right Caputo fractional derivatives holds:*

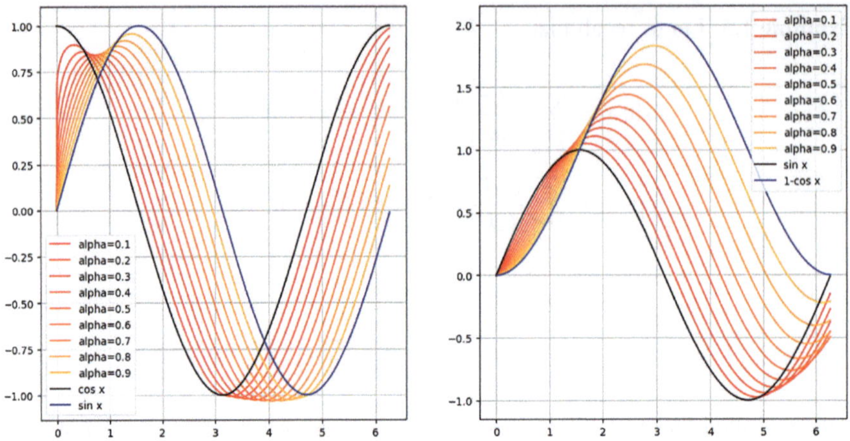

Fig. 1.1 Some left fractional integrals of $\sin x$ (on the right) and $\cos x$ (on the left) with lower limit 0 to illustrate Example 1.3, with integration orders between 0 and 1. An approximation of 175 terms for the Mittag-Leffler function was used

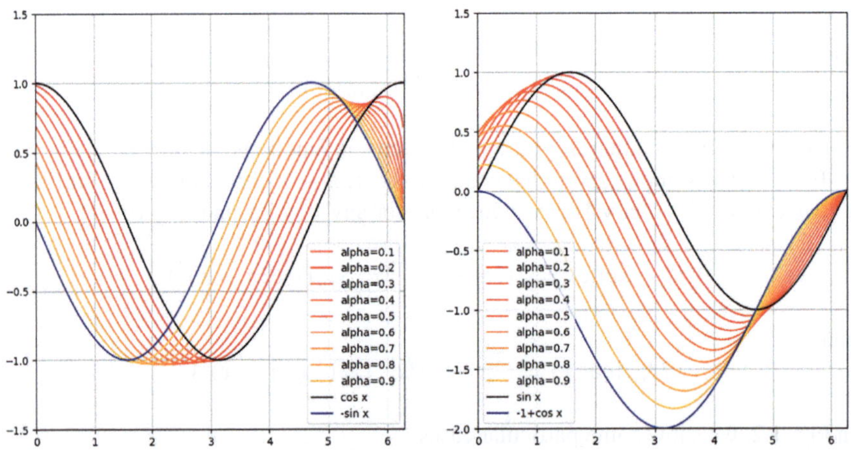

Fig. 1.2 Some right fractional integrals of $\sin x$ (on the right) and $\cos x$ (on the left) with upper limit 2π to illustrate Example 1.3, with integration orders between 0 and 1. An approximation of 175 terms for the Mittag-Leffler function was used

$$D^{\alpha}_{b-} f(x) = Q D^{\alpha}_{a+} Q f(x). \tag{1.8}$$

Proof It is easy to see that $(Qf)^{(n)}(x) = (-1)^n Qf^{(n)}(x)$. Hence, by definition,

$$\begin{aligned} QD^{\alpha}_{a+}Qf(x) &= QI^{n-\alpha}_{a+}(Qf(x))^{(n)} \\ &= (-1)^n QI^{n-\alpha}_{a+}Qf^{(n)}(x) \\ &= (-1)^n I^{n-\alpha}_{b-}f^{(n)}(x) \\ &= D^{\alpha}_{b-}f(x). \end{aligned}$$

□

Remark 1.1 An immediate fact resulting from Definition 1.2 is that, for $\alpha = n \in \mathbb{N}$,

$$D^n_{a+}f(x) = f^{(n)}(x) \text{ and } D^n_{b-}f(x) = (-1)^n f^{(n)}(x). \tag{1.9}$$

Remark 1.2 For the Caputo fractional derivatives to exist almost everywhere on $[a, b]$, it is enough for f to be absolutely continuous of order n on $[a, b]$ ($f \in AC^n([a, b])$),[1] where n is considered as in Definition 1.2 (see Theorem 2.1 from [14]). It is useful to know that

$$C^n([a, b]) \subseteq AC^n([a, b]) \subseteq C^{n-1}([a, b]).$$

In favor of practicity, we will limit ourselves to work with the classes $C^k([a, b])$ for convenient $k \in \mathbb{N}$, partly because of the continuity of the k-th derivative, which allows the Caputo left and right derivatives to exist and be continuous everywhere on $[a, b]$. In addition, such functions offer enough smoothness to attain future results.

Proposition 1.2 *Let $\alpha > 0$, $\alpha \notin \mathbb{N}$ and $n \in \mathbb{N}$ such that $n - 1 < \alpha < n$. If $f \in C^n([a, b])$, then $D^{\alpha}_{a+}f$, $D^{\alpha}_{b-}f \in C([a, b])$ and $D^{\alpha}_{a+}f(a) = D^{\alpha}_{b-}f(b) = 0$.*

Proof Its immediate from Lemma 3.11 from [13], combined with Corollary 1.3. □

Lemma 1.1 *If $f \in C^n([a, b])$ for some $n \in \mathbb{N}$, then there exists $\varphi \in C([a, b])$ such that f can be expressed as*

$$f(x) = I^n_{a+}\varphi(x) + \sum_{k=0}^{n-1} \frac{f^{(k)}(a)}{k!}(x-a)^k \tag{1.10}$$

for any $x \in [a, b]$ and $f^{(n)} = \varphi$.

Proof This is just Taylor's formula. A proof can be found, for example, in Theorem 7.6 from [16]. □

---

[1] We say that f is absolutely continuous in $[a, b]$ if for every $\epsilon > 0$ there exists a $\delta > 0$ such that for any finite set of subintervals $[a_k, b_k]$ of $[a, b]$ s.t. $\sum_k(b_k - a_k) < \delta$, it follows that $\sum_k |f(b_k) - f(a_k)| < \epsilon$. The set of absolutely continuous functions on $[a, b]$ is denoted as $AC([a, b])$. Then we say that $f \in AC^n([a, b])$ if $f \in C^{n-1}([a, b])$ and $f^{(n)} \in AC([a, b])$. A nicer characterization of these functions is given by Theorem 11, p. 126 and Lemma 2.4 from [12, 15], respectively.

Corollary 1.4 *If $f \in C^n([a,b])$, there exists $\varphi \in C([a,b])$ such that*

$$f(x) = (-1)^n I_{b-}^n \varphi(x) + \sum_{k=0}^{n-1} \frac{(-1)^k f^{(k)}(b)}{k!}(b-x)^k \tag{1.11}$$

for any $x \in [a,b]$ and $f^{(n)} = \varphi$.

Proof It follows from factoring a $(-1)^n$ and reversing the integration order in the n-th integral from (1.10). □

The following result will be convenient for the characterization of some special curves.

Lemma 1.2 *Let $\alpha > 0$ and $n \in \mathbb{N}$ such that $n-1 < \alpha \le n$. If $f \in C^n([a,b])$ and $f^{(k)}(a) = 0$ for every $k = 0, \ldots, n-1$, there exists $g \in C^n([a,b])$ such that*

$$f(x) = I_{a+}^{\alpha} g^{(n)}(x) \tag{1.12}$$

for any $x \in [a,b]$. In a similar fashion, if $f^{(k)}(b) = 0$ for $k = 0, \ldots, n-1$, there exists $\tilde{g} \in C^n([a,b])$ such that

$$f(x) = I_{b-}^{\alpha} \tilde{g}^{(n)}(x) \tag{1.13}$$

for any $x \in [a,b]$.

Proof By Lemma 1.1, we can express f as in (1.10). Moreover, due to the semigroup property and Example 1.1, f is represented as

$$f(x) = I_{a+}^{\alpha}\left(I_{a+}^{n-\alpha}\varphi(x) + \sum_{k=0}^{n-1} \frac{f^{(k)}(a)}{\Gamma(k-\alpha+1)}(x-a)^{k-\alpha}\right).$$

Now let

$$h(x) := I_{a+}^{n-\alpha}\varphi(x) + \sum_{k=0}^{n-1} \frac{f^{(k)}(a)}{\Gamma(k-\alpha+1)}(x-a)^{k-\alpha}$$

for every $x \in [a,b]$. Again, using the semigroup property, Example 1.1 and Theorem 1.3, we can rewrite h as

$$\begin{aligned} h(x) &= \frac{d^n}{dx^n}\left(I_{a+}^{2n-\alpha}\varphi(x) + \sum_{k=0}^{n-1} \frac{f^{(k)}(a)}{\Gamma(k+n-\alpha+1)}(x-a)^{k+n-\alpha}\right) \\ &= \frac{d^n}{dx^n}\left(I_{a+}^{2n-\alpha}\varphi(x)\right). \end{aligned}$$

If we define

$$g(x) := I_{a+}^{2n-\alpha}\varphi(x),$$

then g belongs to the class $C^n([a, b])$ and Eq. (1.12) holds.

The right identity (1.13) is proven analogously using Corollary 1.4 and the corresponding semigroup law for right integrals. □

The linearity over the corresponding function spaces is immediate from Definitions 1.1 and 1.2, but more can be said about the relation between the Riemann-Liouville fractional integral and the Caputo fractional derivative. The following result proved in Lemmas 2.21a and 2.22 from [14] constitutes a generalization of the Fundamental Theorem of Calculus.

Theorem 1.3 *Let $f : [a, b] \to \mathbb{R}$ and $\alpha \geq 0$.*

1. *If $f \in C([a, b])$, then*

$$D_{a+}^{\alpha} I_{a+}^{\alpha} f(x) = f(x) \text{ and } D_{b-}^{\alpha} I_{b-}^{\alpha} f(x) = f(x). \tag{1.14}$$

2. *For $f \in C^n([a, b])$, the following identities hold everywhere on $[a, b]$:*

$$I_{a+}^{\alpha} D_{a+}^{\alpha} f(x) = f(x) - \sum_{k=0}^{n-1} \frac{f^{(k)}(a)}{k!}(x-a)^k \text{ and} \tag{1.15}$$

$$I_{b-}^{\alpha} D_{b-}^{\alpha} f(x) = f(x) - \sum_{k=0}^{n-1} \frac{(-1)^k f^{(k)}(b)}{k!}(b-x)^k. \tag{1.16}$$

Example 1.4 As we did for Example 1.1, let us consider $f(x) = (x-a)^{\beta}$ with $\beta \geq 0$, $\alpha > 0$ and $n \in \mathbb{N}$ such that $n - 1 < \alpha \leq n$. Then it follows from Example 3.1 in [13] that

$$D_{a+}^{\alpha} f(x) = \begin{cases} \frac{\Gamma(\beta+1)}{\Gamma(\beta-\alpha+1)}(x-a)^{\beta-\alpha} & \text{if } \beta \notin \{0, 1, \ldots, n-1\} \\ 0 & \text{if } \beta \in \{0, 1, \ldots, n-1\}. \end{cases}$$

Example 1.5 As a consequence of Example 1.4 and Corollary 1.3 applied together, for $f(x) = (b-x)^{\beta}$ with $\beta \geq 0$ we have that

$$D_{b-}^{\alpha} f(x) = \begin{cases} \frac{\Gamma(\beta+1)}{\Gamma(\beta-\alpha+1)}(b-x)^{\beta-\alpha} & \text{if } \beta \notin \{0, 1, \ldots, n-1\} \\ 0 & \text{if } \beta \in \{0, 1, \ldots, n-1\}. \end{cases}$$

It will be relevant for us to know information about the smoothness of the left and right Caputo fractional derivatives. In this sense, the following result proved in Theorem 3.15 from [13] will be useful.

Theorem 1.4 *Let $f \in C^p([a,b])$ for some $p \in \mathbb{N}$ and let $0 < \beta < p$. Then*

$$D_{a+}^{\beta} f(x) = \sum_{r=0}^{p-\lceil\beta\rceil-1} \frac{f^{(r+\lceil\beta\rceil)}(a)}{\Gamma(\lceil\beta\rceil - \beta + r + 1)}(x-a)^{\lceil\beta\rceil-\beta+r} + g(x), \tag{1.17}$$

where $g \in C^{p-\lceil\beta\rceil}([a,b])$ and the $(p-\lceil\beta\rceil)$-th derivative of g satisfies a Lipschitz condition of order $\lceil\beta\rceil - \beta$.

The *right* version of Theorem 1.4, as far as we know, has not been published and for the sake of completeness, we provide a proof here.

Theorem 1.5 *Let $f \in C^p([a,b])$ for some $p \in \mathbb{N}$ and let $0 < \beta < p$. Then*

$$D_{b-}^{\beta} f(x) = \sum_{r=0}^{p-\lceil\beta\rceil-1} \frac{(-1)^{r+\lceil\beta\rceil} f^{(r+\lceil\beta\rceil)}(b)}{\Gamma(\lceil\beta\rceil - \beta + r + 1)}(b-x)^{\lceil\beta\rceil-\beta+r} + g(x), \tag{1.18}$$

where $g \in C^{p-\lceil\beta\rceil}([a,b])$ and the $(p-\lceil\beta\rceil)$-th derivative of g satisfies a Lipschitz condition of order $\lceil\beta\rceil - \beta$.

Proof By Corollary 1.3, $D_{b-}^{\beta} f(x) = Q D_{a+}^{\beta} Q f(x)$ and by Theorem 1.4,

$$\begin{aligned}
D_{b-}^{\beta} f(x) &= Q\left(\sum_{r=0}^{p-\lceil\beta\rceil-1} \frac{(Qf)^{(r+\lceil\beta\rceil)}(a)}{\Gamma(\lceil\beta\rceil - \beta + r + 1)}(x-a)^{\lceil\beta\rceil-\beta+r} + g(x)\right) \\
&= Q\left(\sum_{r=0}^{p-\lceil\beta\rceil-1} \frac{(-1)^{r+\lceil\beta\rceil} f^{(r+\lceil\beta\rceil)}(b)}{\Gamma(\lceil\beta\rceil - \beta + r + 1)}(x-a)^{\lceil\beta\rceil-\beta+r} + g(x)\right) \\
&= \sum_{r=0}^{p-\lceil\beta\rceil-1} \frac{(-1)^{r+\lceil\beta\rceil} f^{(r+\lceil\beta\rceil)}(b)}{\Gamma(\lceil\beta\rceil - \beta + r + 1)}(b-x)^{\lceil\beta\rceil-\beta+r} + g(a+b-x)
\end{aligned}$$

and the theorem follows because $\tilde{g}(x) := g(a+b-x)$ satisfies the same conditions as g. □

As a useful consequence of Theorems 1.4 and 1.5, we have the following

Corollary 1.5 *Let $\alpha > 0$ and $n \in \mathbb{N}$ such that $n-1 < \alpha \le n$. If $p \in \mathbb{N}$ with $p > n$ and $f \in C^p([a,b])$, then $D_{a+}^{\alpha} f, D_{b-}^{\alpha} f \in C^{p-n}((a,b))$.*

An analogous result to Theorem 1.2 holds for Caputo fractional derivatives.

Theorem 1.6 *Let $\alpha \ge 0$, $n \in \mathbb{N}$ with $n-1 < \alpha \le n$ and $\{f_k\}_{k=1}^{\infty}$ be a sequence of functions, such that for each $k \in \mathbb{N}$ we have $f_k \in C^n([a,b])$ and $\{f_k^{(n)}\}_{k=1}^{\infty}$ converges uniformly on $[a,b]$. Then it holds that*

$$D_{a+}^{\alpha} f(x) = \lim_{k\to\infty} D_{a+}^{\alpha} f_k(x)$$

and the sequence $\{D_{a+}^{\alpha} f_k\}$ is uniformly convergent too.

Proof This is a direct consequence of Theorem 1.2 combined with Definition 1.2. □

Corollary 1.6 *Let $\alpha \geq 0$, $n \in \mathbb{N}$ with $n-1 < \alpha \leq n$ and $\{f_k\}_{k=1}^{\infty}$ be a sequence of functions, such that for each $k \in \mathbb{N}$ we have $f_k \in C^n([a,b])$ and $\{f_k^{(n)}\}_{k=1}^{\infty}$ converges uniformly on $[a,b]$. Then it holds that*

$$D_{b-}^{\alpha} f(x) = \lim_{k\to\infty} D_{b-}^{\alpha} f_k(x)$$

and the sequence $\{D_{b-}^{\alpha} f_k\}_{k=1}^{\infty}$ is uniformly convergent too.

Proof As it happened with Corollary 1.1, this result follows from applying Corollary 1.3 to Theorem 1.6. □

Just as we obtained Corollary 1.2, an identical fact follows for Caputo fractional derivatives, by means of considering Theorem 1.6, Corollary 1.6 and Examples 1.4 and 1.5.

Corollary 1.7 *Let $\alpha \geq 0$ and $n \in \mathbb{N}$ with $n-1 < \alpha \leq n$. If f can be represented as a power series within the convergence interval $(a-R, a+R)$ for some convergence radius $R \in (0,\infty]$, let's say $f(x) = \sum_{k=0}^{\infty} \frac{f^{(k)}(a)}{k!}(x-a)^k$, then for any $x \in [a, a+R)$,*

$$D_{a+}^{\alpha} f(x) = \sum_{k=n}^{\infty} \frac{f^{(k)}(a)}{\Gamma(k-\alpha+1)}(x-a)^{k-\alpha}.$$

Analogously, if $f(x) = \sum_{k=0}^{\infty} \frac{f^{(k)}(b)}{k!}(x-b)^k$ within $(b-R, b+R)$, for any $x \in (b-R, b]$ it will follow that

$$D_{b-}^{\alpha} f(x) = \sum_{k=n}^{\infty} \frac{(-1)^k f^{(k)}(b)}{\Gamma(k-\alpha+1)}(b-x)^{k-\alpha}.$$

Example 1.6 Just as we computed expressions for the left fractional integrals of $\sin x$ and $\cos x$, we want to calculate the left Caputo fractional derivatives of such functions, with $a = 0$. If we recall the series representations of $\sin x$ and $\cos x$ from Example 1.3, using Corollary 1.7 we find that for any $x > 0$ and $\alpha \in (0, 1]$,

$$D_{0+}^{\alpha}(\cos x) = x^{-\alpha} E_{2,1-\alpha}(-x^2) - \frac{1}{\Gamma(1-\alpha)} x^{-\alpha}$$

and

$$D_{0+}^{\alpha}(\sin x) = x^{1-\alpha} E_{2,2-\alpha}(-x^2).$$

An approximate plot of some fractional derivatives of these functions is featured in Fig. 1.3.

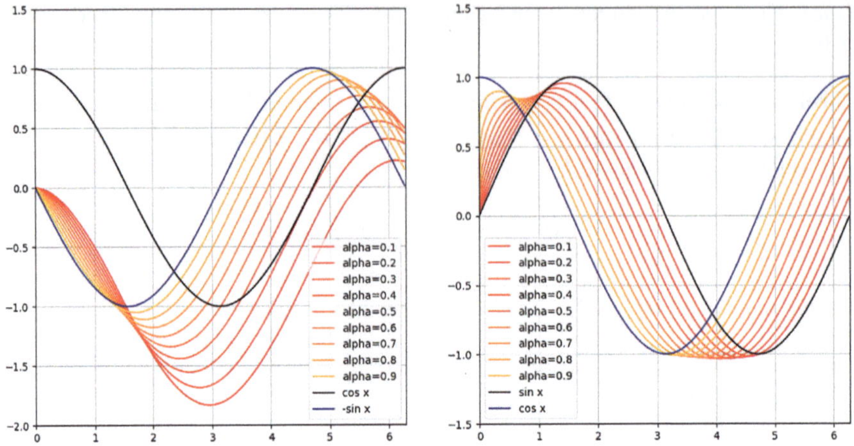

Fig. 1.3 Some Caputo left fractional derivatives of the trigonometric functions for Example 1.6. As in Fig. 1.1, an approximation of 175 terms for the Mittag-Leffler functions was used

1.2 α-Regular Curves and Affine Concepts

Throughout this section, we extend some basic notions of regular curves immersed in the Euclidean space for positive real orders.

For the sake of briefness, in the rest of the paper we denote

$$\boldsymbol{\alpha} := (\alpha_1, \ldots, \alpha_n) \in \mathbb{R}^n,$$

with $\alpha_i > 0$ for every $i = 1, \ldots, n$, unless anything else is specified. In particular, when $\boldsymbol{\alpha} = n(1, \ldots, 1)$ and $n \in \mathbb{N}$, we denote $\boldsymbol{\alpha} =: \mathbf{n}$. In addition, for each $i = 1, \ldots, n$, we let $m_i \in \mathbb{N}$ be a natural number that satisfies $m_i - 1 < \alpha_i \leq m_i$, $\boldsymbol{\mu} := (m_1, \ldots, m_n)$ and $m := \max_{1 \leq i \leq n}\{m_i\}$.

Firstly, we need a new notion of the classical tangent vector to a curve, in order to extract information about its extrinsic geometry in $\mathbb{R}^n$, in a fractional sense. The following definition is based on the notion given in [11] for plane curves and here we extend it admitting the derivation orders to be arbitrary for each component of the curve.

Remark 1.3 Throughout the rest of the article, left and right fractional generalizations of classical concepts of the theory of curves in $\mathbb{R}^n$ will be introduced, and as we will see later, these concepts may coincide for integer orders (See Remarks 1.5, 1.6, 1.8, 1.9 and 1.10). Therefore, whenever this coincidence takes place, we omit the directional specification of the considered concept.

Definition 1.3 Let $\gamma \in C^m([a,b],\mathbb{R}^n)$. We define the **left $\boldsymbol{\alpha}$-tangent vector to γ at t** as

$$D^{\boldsymbol{\alpha}}_{a+}\gamma(t) := (D^{\alpha_1}_{a+}\gamma_1(t), \ldots, D^{\alpha_n}_{a+}\gamma_n(t)). \tag{1.19}$$

Analogously, the **right $\boldsymbol{\alpha}$-tangent vector to γ at t** is expressed as

$$D^{\boldsymbol{\alpha}}_{b-}\gamma(t) := (D^{\alpha_1}_{b-}\gamma_1(t), \ldots, D^{\alpha_n}_{b-}\gamma_n(t)). \tag{1.20}$$

In the same spirit, we have the following.

Definition 1.4 For $\gamma \in C([a,b],\mathbb{R}^n)$, the **left $\boldsymbol{\alpha}$-integral vector to γ at t** is given by the vector

$$I^{\boldsymbol{\alpha}}_{a+}\gamma(t) := (I^{\alpha_1}_{a+}\gamma_1(t), \ldots, I^{\alpha_n}_{a+}\gamma_n(t)) \tag{1.21}$$

and the **right $\boldsymbol{\alpha}$-integral vector to γ at t is given as**

$$I^{\boldsymbol{\alpha}}_{b-}\gamma(t) := (I^{\alpha_1}_{b-}\gamma_1(t), \ldots, I^{\alpha_n}_{b-}\gamma_n(t)). \tag{1.22}$$

A fractional notion of regularity is given next, using Definition 1.3.

Definition 1.5 Let $\gamma \in C^m([a,b],\mathbb{R}^n)$. We say that γ is a **left $\boldsymbol{\alpha}$-regular curve at t** if $D^{\boldsymbol{\alpha}}_{a+}\gamma(t) \neq \mathbf{0}$. In particular, if $\boldsymbol{\alpha} = \alpha(1,\ldots,1)$, we'll just say that γ is **left α-regular at t**.

Similarly, γ will be called **a right $\boldsymbol{\alpha}$-regular curve at t** if $D^{\boldsymbol{\alpha}}_{b-}\gamma(t) \neq \mathbf{0}$. Again, if $\boldsymbol{\alpha} = \alpha(1,\ldots,1)$, we say that γ is **right α-regular at t**.

In addition, we say that γ is **left (resp. right)-$\boldsymbol{\alpha}$-regular in $U \subseteq (a,b)$** if it is left (resp. right) $\boldsymbol{\alpha}$-regular for every $t \in U$, and if $U = (a,b)$ or U is clear from the context, we just say that γ is left (resp. right) $\boldsymbol{\alpha}$-regular.

Remark 1.4 The terminals a and b are excluded in Definition 1.5 due to Proposition 1.2.

Remark 1.5 When $\boldsymbol{\alpha} = (m_1,\ldots,m_n)$, the notions of left and right regularity coincide as a consequence from Remark 1.1: in the left case, the left $\boldsymbol{\alpha}$-tangent vector to γ reduces to

$$D^{\boldsymbol{\alpha}}_{a+}\gamma(t) = (\gamma_1^{(m_1)}(t), \ldots, \gamma_n^{(m_n)}(t)),$$

whereas for the right case,

$$D^{\boldsymbol{\alpha}}_{b-}\gamma(t) = ((-1)^{m_1}\gamma_1^{(m_1)}(t), \ldots, (-1)^{m_n}\gamma_n^{(m_n)}(t)).$$

Now γ isn't left $\boldsymbol{\alpha}$-regular in $U \subseteq (a,b)$ if and only if $\gamma_i^{(m_i)}(t) = 0$ for some $t \in U$, if and only if $(-1)^{m_i}\gamma_i^{(m_i)}(t) = 0$ for some $t \in U$, which is equivalent to γ not being right $\boldsymbol{\alpha}$-regular in U.

Let us introduce some notation.

Definition 1.6 Let us express the **set of left $\boldsymbol{\alpha}$-regular curves in** $U \subseteq (a, b)$ as

$$\mathcal{R}_{a+}(\boldsymbol{\alpha}; U, \mathbb{R}^n) := \{\gamma \in C^m([a, b], \mathbb{R}^n) : \gamma \text{ is left } \boldsymbol{\alpha}\text{-regular in } U\}. \tag{1.23}$$

In a similar manner the **set of right $\boldsymbol{\alpha}$-regular curves in** $U \subseteq (a, b)$ **is defined as**

$$\mathcal{R}_{b-}(\boldsymbol{\alpha}; U, \mathbb{R}^n) := \{\gamma \in C^m([a, b], \mathbb{R}^n) : \gamma \text{ is right } \boldsymbol{\alpha}\text{-regular in } U\}. \tag{1.24}$$

As in Definition 1.5, if $\boldsymbol{\alpha} = \alpha(1, \ldots, 1)$, we let $\mathcal{R}_{a+}(\boldsymbol{\alpha}; U, \mathbb{R}^n) =: \mathcal{R}_{a+}(\alpha; U, \mathbb{R}^n)$ and $\mathcal{R}_{b-}(\boldsymbol{\alpha}; U, \mathbb{R}^n) =: \mathcal{R}_{b-}(\alpha; U, \mathbb{R}^n)$.

Remark 1.6 Due to Remark 1.5, when $\boldsymbol{\alpha} \in \mathbb{N}^n$, we will drop the left and right specification from the $\boldsymbol{\alpha}$-regularity and with regard to Definition 1.6, we set

$$\mathcal{R}_{a+}(\boldsymbol{\alpha}; U, \mathbb{R}^n) = \mathcal{R}_{b-}(\boldsymbol{\alpha}; U, \mathbb{R}^n) = \mathcal{R}(\boldsymbol{\alpha}; U, \mathbb{R}^n).$$

Remark 1.7 A particular but noticeable case is the set of regular curves in (a, b), in the classical sense, which equals $\mathcal{R}(1; (a, b), \mathbb{R}^n)$ under the notation of Definition 1.6.

Example 1.7 For $\boldsymbol{\alpha} \in (0, 1]^n \subseteq \mathbb{R}^n$, a wide family of left (resp. right) $\boldsymbol{\alpha}$-regular curves in $U \subseteq (a, b)$ is given by the sets

$$D_{a+}^{\mathbf{1}-\boldsymbol{\alpha}}(\mathcal{R}(1; U, \mathbb{R}^n)) = \{\gamma : \exists\, \tilde{\gamma} \in \mathcal{R}(1; U, \mathbb{R}^n),\ \gamma(t) = D_{a+}^{\mathbf{1}-\boldsymbol{\alpha}}\tilde{\gamma}(t)\}$$

and

$$D_{b-}^{\mathbf{1}-\boldsymbol{\alpha}}(\mathcal{R}(1; U, \mathbb{R}^n)) = \{\gamma : \exists\, \tilde{\gamma} \in \mathcal{R}(1; U, \mathbb{R}^n),\ \gamma(t) = D_{b-}^{\mathbf{1}-\boldsymbol{\alpha}}\tilde{\gamma}(t)\},$$

respectively. This follows from Theorem 1.3, because if $\gamma \in D_{a+}^{\mathbf{1}-\boldsymbol{\alpha}}(\mathcal{R}_{a+}(1; U, \mathbb{R}^n))$, $\gamma(t) = D_{a+}^{\mathbf{1}-\boldsymbol{\alpha}}\tilde{\gamma}(t)$ and for each $i = 1, \ldots, n$,

$$\begin{aligned} D_{a+}^{\alpha_i}\gamma_i(t) &= D_{a+}^{\alpha_i} D_{a+}^{1-\alpha_i}\tilde{\gamma}_i(t) \\ &= D_{a+}^{\alpha_i} I_{a+}^{\alpha_i}\tilde{\gamma}_i'(t) \\ &= \tilde{\gamma}_i'(t) \end{aligned}$$

and the reasoning for the right case is identical. More generally, for $\boldsymbol{\alpha} \in (0, \infty)^n$, the families

$$L_{a+}^{\boldsymbol{\alpha}}(\mathcal{R}(\boldsymbol{\mu}; U, \mathbb{R}^n)) = \{\gamma : \exists\, \tilde{\gamma} \in \mathcal{R}(\boldsymbol{\mu}; U, \mathbb{R}^n),\ \gamma(t) = L_{a+}^{\boldsymbol{\alpha}}\tilde{\gamma}(t)\}$$

and

$$L_{b-}^{\boldsymbol{\alpha}}(\mathcal{R}(\boldsymbol{\mu}; U, \mathbb{R}^n)) = \{\gamma : \exists\, \tilde{\gamma} \in \mathcal{R}(\boldsymbol{\mu}; U, \mathbb{R}^n),\ \gamma(t) = L_{b-}^{\boldsymbol{\alpha}}\tilde{\gamma}(t)\}$$

are examples of left and right $\boldsymbol{\alpha}$-regular curves in $U \subseteq (a, b)$, respectively, where we have defined the left and right linear operators $L^{\boldsymbol{\alpha}}_{a+}\gamma := I^{\boldsymbol{\alpha}}_{a+}D^{\boldsymbol{\mu}}_{a+}\gamma$ and $L^{\boldsymbol{\alpha}}_{b-}\gamma := I^{\boldsymbol{\alpha}}_{b-}D^{\boldsymbol{\mu}}_{b-}\gamma$.

Even more can be said about the curves of Example 1.7, by means of Lemma 1.2.

Theorem 1.7 *Let $\gamma \in \mathcal{R}_{a+}(\boldsymbol{\alpha}; (a, b), \mathbb{R}^n)$ (resp. $\gamma \in \mathcal{R}_{b-}(\boldsymbol{\alpha}; (a, b), \mathbb{R}^n)$). If*

$$\gamma^{(k)}(a) = \mathbf{0} \text{ (Resp. } \gamma^{(k)}(b) = \mathbf{0}), \quad k = 0, 1, \ldots, m-1,$$

then $\gamma \in L^{\boldsymbol{\alpha}}_{a+}(\mathcal{R}(\boldsymbol{\mu}; (a, b), \mathbb{R}^n))$ (resp. $\gamma \in L^{\boldsymbol{\alpha}}_{b-}(\mathcal{R}(\boldsymbol{\mu}; (a, b), \mathbb{R}^n))$.

Proof We only prove the left case, because the right case is identical. In view that the components of the curve γ satisfy the hypotheses of Lemma 1.2, for each component γ_i we can find some $\tilde{\gamma}_i \in C^m([a, b])$ such that $\gamma_i(t) = I^{\alpha_i}_{a+}\tilde{\gamma}_i^{(m_i)}(t)$. Hence, there exists some $\tilde{\gamma} \in C^m([a, b], \mathbb{R}^n)$ such that $\gamma = I^{\boldsymbol{\alpha}}_{a+}D^{\boldsymbol{\mu}}_{a+}\tilde{\gamma}$. Now, if $\tilde{\gamma}$ were not $\boldsymbol{\mu}$-regular, there would exist some $t \in (a, b)$ with $D^{\boldsymbol{\mu}}_{a+}\tilde{\gamma}(t) = \mathbf{0}$, which would imply that $D^{\boldsymbol{\alpha}}_{a+}\gamma(t) = \mathbf{0}$, contradicting our hypothesis. □

Analogously to the classical arc length notion, we introduce the following concept.

Definition 1.7 If $\gamma \in C^m([a, b], \mathbb{R}^n)$, we define its **left $\boldsymbol{\alpha}$-arc length** throughout the following integral:

$$s^{\boldsymbol{\alpha}}_{a+}(t) := \int_a^t ||D^{\boldsymbol{\alpha}}_{a+}\gamma(\tau)||d\tau. \tag{1.25}$$

Similarly, the **right $\boldsymbol{\alpha}$-arc length of γ is given as**

$$s^{\boldsymbol{\alpha}}_{b-}(t) := -\int_t^b ||D^{\boldsymbol{\alpha}}_{b-}\gamma(\tau)||d\tau. \tag{1.26}$$

The left (resp. right) $\boldsymbol{\alpha}$-regularity turns out to be a sufficient condition for the left (resp. right) $\boldsymbol{\alpha}$-arc length to be a local diffeomorfism.

Proposition 1.3 *If $\gamma \in C^m([a, b], \mathbb{R}^n)$ is a left (resp. right) $\boldsymbol{\alpha}$-regular curve, then $s^{\boldsymbol{\alpha}}_{a+}$ (resp. $s^{\boldsymbol{\alpha}}_{b-}$) is a diffeomorphism from (a, b) to $s^{\boldsymbol{\alpha}}_{a+}((a, b))$ (resp. $s^{\boldsymbol{\alpha}}_{b-}((a, b))$).*

Proof For the left case, due to the First Fundamental Theorem of Calculus,

$$\frac{ds^{\boldsymbol{\alpha}}_{a+}}{dt} = ||D^{\boldsymbol{\alpha}}_{a+}\gamma(t)|| > 0,$$

where the last inequality follows from the hypothesis of left $\boldsymbol{\alpha}$-regularity. Finally, we can conclude that $s^{\boldsymbol{\alpha}}_{a+}$ is invertible in (a, b) with differentiable inverse $(s^{\boldsymbol{\alpha}}_{a+})^{-1}$ in $s^{\boldsymbol{\alpha}}_{a+}((a, b))$, by the inverse function theorem in one variable. The proof for the right case is analogous. □

In other words, Proposition 1.3 implies that a left (resp. right) $\boldsymbol{\alpha}$-regular curve can be parametrized by its left (resp. right) $\boldsymbol{\alpha}$-arc length. In connection with this fact, we have the following.

Definition 1.8 Let $\gamma \in C^m([a, b], \mathbb{R}^n)$. We say that γ is **left $\boldsymbol{\alpha}$-unit speed** if for each $t \in (a, b)$ we have

$$||D^{\boldsymbol{\alpha}}_{a+}\gamma(t)|| = 1. \tag{1.27}$$

Analogously, we say that γ is **right $\boldsymbol{\alpha}$-unit speed** if for each $t \in (a, b)$ it follows that

$$||D^{\boldsymbol{\alpha}}_{b-}\gamma(t)|| = 1. \tag{1.28}$$

Remark 1.8 In the same spirit as Definitions 1.5 and 1.6, for $\boldsymbol{\alpha} = \alpha(1, \ldots, 1)$ we talk about left and right α-arc length and, correspondingly, left and right α-unit speed. Besides, the left and right notions proposed in Definitions 1.7 and 1.8 are equivalent for $\boldsymbol{\alpha} \in \mathbb{N}^n$, in the same way as in Remark 1.5.

Proposition 1.4 *Let $\gamma \in C^m([a, b], \mathbb{R}^n)$ be an $\boldsymbol{\alpha}$-regular curve. Then γ is left (resp. right) $\boldsymbol{\alpha}$-unit speed if and only if γ is parametrized (up to translation) by left (resp. right) $\boldsymbol{\alpha}$-arc length.*

Proof We prove the left case and we leave the similar details for the right case as an exercise.

If $||D^{\boldsymbol{\alpha}}_{a+}\gamma(t)|| = 1$ for each $t \in (a, b)$, its immediate from Definition 1.7 that

$$s^{\boldsymbol{\alpha}}(t) = t - a.$$

Hence, up to translation by a, we can conclude that the parameter t equals the left $\boldsymbol{\alpha}$-arc length $s^{\boldsymbol{\alpha}}_{a+}$.

Inversely, if $s^{\boldsymbol{\alpha}}_{a+}(t) = t - t_0$, we have $\dfrac{ds^{\boldsymbol{\alpha}}_{a+}}{dt} = 1 = ||D^{\alpha}_{a+}\gamma(t)||$ for each $t \in (a, b)$ and γ is left $\boldsymbol{\alpha}$-unit speed.

□

Example 1.8 For $\boldsymbol{\alpha} \in (0, 1]^n$ and any $\gamma \in D^{\mathbf{1}-\boldsymbol{\alpha}}_{a+}(\mathcal{R}(1; (a, b), \mathbb{R}^n))$, we find that the computation of the left $\boldsymbol{\alpha}$-arc length amounts to calculate a 1-arc length, because $\gamma(t) = D^{\mathbf{1}-\boldsymbol{\alpha}}_{a+}\tilde{\gamma}(t)$ for some 1-regular curve $\tilde{\gamma} \in C^1([a, b], \mathbb{R}^n)$ and

$$\begin{aligned} s^{\boldsymbol{\alpha}}_{a+}(t) &= \int_a^t ||D^{\boldsymbol{\alpha}}_{a+}\gamma(\tau)||d\tau \\ &= \int_a^t ||\tilde{\gamma}'(\tau)||d\tau \\ &= \tilde{s}^1(t). \end{aligned}$$

A similar situation happens with the right case, where we have that

$$
\begin{aligned}
s_{b-}^{\alpha}(t) &= -\int_t^b ||D_{b-}^{\alpha}\gamma(\tau)||d\tau \\
&= -\int_t^b ||(-1)\tilde{\gamma}'(\tau)||d\tau \\
&= -\int_t^b ||\tilde{\gamma}'(\tau)||d\tau \\
&= \tilde{s}^1(t) - \tilde{s}^1(b).
\end{aligned}
$$

1.3 Frenet-Serret Equations of Order α

Aiming to generalize the classical Frenet-Serret differential equations, we are interested in the subclass of left (resp. right) $\boldsymbol{\alpha}$-regular curves whose $n-1$ fractional vectors $\frac{d^i}{dt^i} D_{a+}^{\boldsymbol{\alpha}}\gamma(t)$ (resp. $\frac{d^i}{dt^i} D_{b-}^{\boldsymbol{\alpha}}\gamma(t)$) for $i = 0, \ldots, n-2$ are linearly independent for each t. For the rest of the paper, we let $p := \max\{m, n\}$.

Definition 1.9 We say that $\gamma \in C^{2p}([a,b], \mathbb{R}^n)$ is a **left $\boldsymbol{\alpha}$-Frenet curve** if for every $t \in (a,b)$ it follows that the vectors

$$
D_{a+}^{\boldsymbol{\alpha}}\gamma(t), \frac{d}{dt} D_{a+}^{\boldsymbol{\alpha}}\gamma(t), \ldots, \frac{d^{n-2}}{dt^{n-2}} D_{a+}^{\boldsymbol{\alpha}}\gamma(t)
$$

are linearly independent. Analogously, we say that γ is a **right $\boldsymbol{\alpha}$-Frenet curve** if for every $t \in (a,b)$ we have that the vectors

$$
D_{b-}^{\boldsymbol{\alpha}}\gamma(t), \frac{d}{dt} D_{b-}^{\boldsymbol{\alpha}}\gamma(t), \ldots, \frac{d^{n-2}}{dt^{n-2}} D_{b-}^{\boldsymbol{\alpha}}\gamma(t)
$$

are linearly independent. As we did for Definition 1.6, we let $\mathcal{F}_{a+}(\boldsymbol{\alpha}; (a,b), \mathbb{R}^n)$ and $\mathcal{F}_{b-}(\boldsymbol{\alpha}; (a,b), \mathbb{R}^n)$ be the left and right sets of $\boldsymbol{\alpha}$-Frenet curves, respectively. Following our previous conventions, we talk about **left (resp. right) α-Frenet curves** if $\boldsymbol{\alpha} = \alpha(1, \ldots, 1)$.

Remark 1.9 For $\boldsymbol{\alpha} \in \mathbb{N}^n$, let's say $\boldsymbol{\alpha} = \boldsymbol{\mu}$, it is true that

$$
\mathcal{F}_{a+}(\boldsymbol{\alpha}; (a,b), \mathbb{R}^n) = \mathcal{F}_{b-}(\boldsymbol{\alpha}; (a,b), \mathbb{R}^n).
$$

This equivalence stems from the fact that linear independence is preserved by injective linear mappings. Let $J := \operatorname{diag}((-1)^{m_1}, \ldots, (-1)^{m_n})$. Given that left and right Caputo fractional derivatives coincide up to sign for integer orders, it is true that for each $i = 0, \ldots, n-2$,

$$
\frac{d^i}{dt^i} D_{b-}^{\boldsymbol{\alpha}}\gamma(t) = J\left(\frac{d^i}{dt^i} D_{a+}^{\boldsymbol{\alpha}}\gamma(t)\right).
$$

Hence, the linear independence of the set

$$\left\{D^{\boldsymbol{\alpha}}_{a+}\gamma(t), \frac{d}{dt}D^{\boldsymbol{\alpha}}_{a+}\gamma(t), \ldots, \frac{d^{n-2}}{dt^{n-2}}D^{\boldsymbol{\alpha}}_{a+}\gamma(t)\right\}$$

is equivalent to the linear independence of the set

$$\left\{D^{\boldsymbol{\alpha}}_{b-}\gamma(t), \frac{d}{dt}D^{\boldsymbol{\alpha}}_{b-}\gamma(t), \ldots, \frac{d^{n-2}}{dt^{n-2}}D^{\boldsymbol{\alpha}}_{b-}\gamma(t)\right\}.$$

Therefore, we will drop the subscripts of the sets $\mathcal{F}_{a+}(\boldsymbol{\alpha}; (a, b), \mathbb{R}^n)$ and $\mathcal{F}_{b-}(\boldsymbol{\alpha}; (a, b), \mathbb{R}^n)$ when $\boldsymbol{\alpha} \in \mathbb{N}^n$.

Example 1.9 As it happened with Example 1.7, for $\boldsymbol{\alpha} \in (0, 1]^n$ we obtain a family of left (resp. right) $\boldsymbol{\alpha}$-Frenet curves using 1-Frenet curves:

$$D^{\mathbf{1}-\boldsymbol{\alpha}}_{a+}(\mathcal{F}(1; (a, b), \mathbb{R}^n)) = \{\gamma : \exists\, \tilde{\gamma} \in \mathcal{F}(1; (a, b), \mathbb{R}^n),\ \gamma(t) = D^{\mathbf{1}-\boldsymbol{\alpha}}_{a+}\tilde{\gamma}(t)\}$$

and

$$D^{\mathbf{1}-\boldsymbol{\alpha}}_{b-}(\mathcal{F}(1; (a, b), \mathbb{R}^n)) = \{\gamma : \exists\, \tilde{\gamma} \in \mathcal{F}(1; (a, b), \mathbb{R}^n),\ \gamma(t) = D^{\mathbf{1}-\boldsymbol{\alpha}}_{b-}\tilde{\gamma}(t)\},$$

respectively. More generally, for $\boldsymbol{\alpha} \in (0, \infty)^n$, the following are respective families of left and right $\boldsymbol{\alpha}$-Frenet curves:

$$L^{\boldsymbol{\alpha}}_{a+}(\mathcal{F}(\boldsymbol{\mu}; (a, b), \mathbb{R}^n)) = \{\gamma : \exists\, \tilde{\gamma} \in \mathcal{F}(\boldsymbol{\mu}; (a, b), \mathbb{R}^n),\ \gamma(t) = L^{\boldsymbol{\alpha}}_{a+}\tilde{\gamma}(t)\}$$

and

$$L^{\boldsymbol{\alpha}}_{b-}(\mathcal{F}(\boldsymbol{\mu}; (a, b), \mathbb{R}^n)) = \{\gamma : \exists\, \tilde{\gamma} \in \mathcal{F}(\boldsymbol{\mu}; (a, b), \mathbb{R}^n),\ \gamma(t) = L^{\boldsymbol{\alpha}}_{b-}\tilde{\gamma}(t)\}.$$

The linear independence property of the left (resp. right) $\boldsymbol{\alpha}$-Frenet curves allows us to construct a moving frame which moves alongside with the curve. For such purpose, let us remember that in $\mathbb{R}^n$, the **cross product of the** $n-1$ **vectors** $\mathbf{v}_1, \ldots, \mathbf{v}_{n-1}$ can be defined as the only vector $\mathbf{v} \in \mathbb{R}^n$ such that

$$\langle \mathbf{v}, \mathbf{x} \rangle = \det(\mathbf{x}, \mathbf{v}_1, \ldots, \mathbf{v}_{n-1})$$

for any $\mathbf{x} \in \mathbb{R}^n$. Such vector is denoted as $\mathbf{v}_1 \times \cdots \times \mathbf{v}_{n-1}$. In [17], it is proved that an explicit expression for the cross product is given as

$$\mathbf{v}_1 \times \cdots \times \mathbf{v}_{n-1} = \begin{vmatrix} \mathbf{e}_1 & \cdots & \mathbf{e}_n \\ v_{1,1} & \cdots & v_{1,n} \\ \vdots & \ddots & \vdots \\ v_{n-1,1} & \cdots & v_{n-1,n} \end{vmatrix},$$

where $v_{i,j}$ stands for the j-th component of the i-th vector. In the rest of the article, we will abbreviate the notation as follows:

$$D_{a+}^{i,\alpha}\gamma(t) := \frac{d^i}{dt^i} D_{a+}^{\alpha}\gamma(t) \text{ and } D_{b-}^{i,\alpha}\gamma(t) := \frac{d^i}{dt^i} D_{b-}^{\alpha}\gamma(t).$$

Proposition 1.5 *Let $\gamma \in \mathcal{F}_{a+}(\boldsymbol{\alpha}; (a,b), \mathbb{R}^n)$ (resp. $\gamma \in \mathcal{F}_{b-}(\boldsymbol{\alpha}; (a,b), \mathbb{R}^n)$). For each $t \in (a,b)$, there exists a unique basis $\{\mathbf{u}_1(t), \ldots, \mathbf{u}_n(t)\}$ for $\mathbb{R}^n$ satisfying that*

1. *$\{\mathbf{u}_1(t), \ldots, \mathbf{u}_n(t)\}$ is orthonormal and positively oriented;*
2. *For each $i = 0, \ldots, n-2$, $\text{span}\{D_{a+}^{\alpha}\gamma(t), \ldots, D_{a+}^{i,\alpha}\gamma(t)\} = \text{span}\{\mathbf{u}_1(t), \ldots, \mathbf{u}_{i+1}(t)\}$ (resp. $\text{span}\{D_{b-}^{\alpha}\gamma(t), \ldots, D_{b-}^{i,\alpha}\gamma(t)\} = \text{span}\{\mathbf{u}_1(t), \ldots, \mathbf{u}_{i+1}(t)\}$);*
3. *$\langle D_{a+}^{i,\alpha}\gamma(t), \mathbf{u}_{i+1}(t)\rangle > 0$ (resp. $\langle D_{b-}^{i,\alpha}\gamma(t), \mathbf{u}_{i+1}(t)\rangle > 0$) for each $i = 0, \ldots, n-2$.*

Proof We only provide the proof for the left case; the right case is identical.

For each $t \in (a,b)$, let us consider the linearly independent set $\{D_{a+}^{\alpha}\gamma(t), \ldots, D_{a+}^{n-2,\alpha}\gamma(t)\}$. Let us orthonormalize this set via the Gram-Schmidt algorithm and we will get an orthonormal set $\{\mathbf{u}_1(t), \ldots, \mathbf{u}_{n-1}(t)\}$ as follows:

$$\begin{aligned}
\mathbf{u}_1(t) &:= \frac{D_{a+}^{\alpha}\gamma(t)}{||D_{a+}^{\alpha}\gamma(t)||}, \\
\mathbf{u}_2(t) &:= \frac{D_{a+}^{1,\alpha}\gamma(t) - \langle D_{a+}^{1,\alpha}\gamma(t), \mathbf{u}_1(t)\rangle \mathbf{u}_1(t)}{||D_{a+}^{1,\alpha}\gamma(t) - \langle D_{a+}^{1,\alpha}\gamma(t), \mathbf{u}_1(t)\rangle \mathbf{u}_1(t)||}, \\
&\vdots \\
\mathbf{u}_k(t) &:= \frac{D_{a+}^{k-1,\alpha}\gamma(t) - \sum_{j=1}^{k-1}\langle D_{a+}^{k-1,\alpha}\gamma(t), \mathbf{u}_j(t)\rangle \mathbf{u}_j(t)}{\left|\left|D_{a+}^{k-1,\alpha}\gamma(t) - \sum_{j=1}^{k-1}\langle D_{a+}^{k-1,\alpha}\gamma(t), \mathbf{u}_j(t)\rangle \mathbf{u}_j(t)\right|\right|}, \\
&\vdots \\
\mathbf{u}_{n-1}(t) &:= \frac{D_{a+}^{n-2,\alpha}\gamma(t) - \sum_{j=1}^{n-2}\langle D_{a+}^{n-2,\alpha}\gamma(t), \mathbf{u}_j(t)\rangle \mathbf{u}_j(t)}{\left|\left|D_{a+}^{n-2,\alpha}\gamma(t) - \sum_{j=1}^{n-2}\langle D_{a+}^{n-2,\alpha}\gamma(t), \mathbf{u}_j(t)\rangle \mathbf{u}_j(t)\right|\right|}.
\end{aligned}$$

Moreover, we can complete this set to an orthonormal basis for $\mathbb{R}^n$ by adjoining it the unit vector $\mathbf{u}_n(t) := \mathbf{u}_1(t) \times \cdots \times \mathbf{u}_{n-1}(t)$. Hence, due to the properties of the Gram-Schmidt process, the orthonormal basis $\{\mathbf{u}_1(t), \ldots, \mathbf{u}_n(t)\}$ will fulfill property 2. Besides, properties 1 and 3 will characterize uniquely this basis because of the choice of the signs of the vectors $\mathbf{u}_i(t)$, and the requirement of positive orientation. □

Definition 1.10 The unique moving frame from Proposition 1.5 for $\gamma \in \mathcal{F}_{a+}(\boldsymbol{\alpha}; (a,b), \mathbb{R}^n)$ (resp. $\gamma \in \mathcal{F}_{b-}(\boldsymbol{\alpha}; (a,b), \mathbb{R}^n)$) will be known as its **left $\boldsymbol{\alpha}$-Frenet**

frame (resp. right α-Frenet frame) and following our conventions, the definitions reduce to **left α-Frenet frame (resp. right α-Frenet frame)** when $\boldsymbol{\alpha} = \alpha(1, \ldots, 1)$.

Remark 1.10 Due to Remark 1.9, the left and right $\boldsymbol{\alpha}$-Frenet frames coincide when $\boldsymbol{\alpha} = \boldsymbol{\mu} \in \mathbb{N}^n$.

Remark 1.11 The smoothness of the left (resp. right) $\boldsymbol{\alpha}$- Frenet frame will be relevant for our purposes. Notice that the construction developed in Proposition 1.5, alongside with Corollary 1.5 imply together that $\mathbf{u}_1 \in C^{2p-n}((a,b), \mathbb{R}^n)$. Given that $D^{1,\boldsymbol{\alpha}}_{a+}\gamma \in C^{2p-n-1}((a,b), \mathbb{R}^n)$, we have that $\mathbf{u}_2 \in C^{2p-n-1}((a,b), \mathbb{R}^n)$ and inductively, it follows that $\mathbf{u}_j \in C^{2p-n-j+1}((a,b), \mathbb{R}^n)$ for $j = 1, \ldots, n-1$ and finally, given that $\mathbf{u}_n(t)$ is the cross product of the previous vector functions, it follows that $\mathbf{u}_n \in C^{2(p-n+1)}((a,b), \mathbb{R}^n)$.

Example 1.10 For planar and spatial curves ($n = 2$ and $n = 3$, respectively), the left (resp. right) $\boldsymbol{\alpha}$-Frenet frames adopt explicit and useful forms.

Firstly, for $n = 2$, in fact, any left (resp. right) $\boldsymbol{\alpha}$-regular curve is left (resp. right) $\boldsymbol{\alpha}$-Frenet by Definition 1.9. Hence, if $\gamma \in \mathcal{R}_{a+}(\boldsymbol{\alpha}; (a,b), \mathbb{R}^2)$ (resp. $\gamma \in \mathcal{R}_{b-}(\boldsymbol{\alpha}; (a,b), \mathbb{R}^2)$), let $B := \begin{pmatrix} 0 & -1 \\ 1 & 0 \end{pmatrix}$ and we must have $\mathbf{u}_2(t) = B\mathbf{u}_1(t)$. Explicitly,

$$\mathbf{u}_1(t) = \frac{(D^{\alpha_1}_{a+}\gamma_1(t), D^{\alpha_2}_{a+}\gamma_2(t))}{\sqrt{(D^{\alpha_1}_{a+}\gamma_1(t))^2 + (D^{\alpha_2}_{a+}\gamma_2(t))^2}} \text{ and}$$

$$\mathbf{u}_2(t) = \frac{(-D^{\alpha_2}_{a+}\gamma_2(t), D^{\alpha_1}_{a+}\gamma_1(t))}{\sqrt{(D^{\alpha_1}_{a+}\gamma_1(t))^2 + (D^{\alpha_2}_{a+}\gamma_2(t))^2}}.$$

$$\left(\text{Resp. } \mathbf{u}_1(t) = \frac{(D^{\alpha_1}_{b-}\gamma_1(t), D^{\alpha_2}_{b-}\gamma_2(t))}{\sqrt{(D^{\alpha_1}_{b-}\gamma_1(t))^2 + (D^{\alpha_2}_{b-}\gamma_2(t))^2}} \text{ and}\right.$$

$$\left.\mathbf{u}_2(t) = \frac{(-D^{\alpha_2}_{b-}\gamma_2(t), D^{\alpha_1}_{b-}\gamma_1(t))}{\sqrt{(D^{\alpha_1}_{b-}\gamma_1(t))^2 + (D^{\alpha_2}_{b-}\gamma_2(t))^2}}.\right)$$

This follows because $\mathbf{u}_2(t)$ given as before makes the basis $\{\mathbf{u}_1(t), \mathbf{u}_2(t)\}$ orthonormal and positively oriented.

For $n = 3$, we already know that $\mathbf{u}_1(t) = \frac{D^{\alpha}_{a+}\gamma(t)}{||D^{\alpha}_{a+}\gamma(t)||}$ (resp. $\mathbf{u}_1(t) = \frac{D^{\alpha}_{b-}\gamma(t)}{||D^{\alpha}_{b-}\gamma(t)||}$) and the second vector of the left (resp. right) $\boldsymbol{\alpha}$-Frenet frame can be expressed as

$$\mathbf{u}_2(t) = \frac{D^{\alpha}_{a+}\gamma(t) \times (D^{1,\alpha}_{a+}\gamma(t) \times D^{\alpha}_{a+}\gamma(t))}{||D^{\alpha}_{a+}\gamma(t) \times (D^{1,\alpha}_{a+}\gamma(t) \times D^{\alpha}_{a+}\gamma(t))||}$$

$$\left(\text{Respectively, } \mathbf{u}_2(t) = \frac{D^{\alpha}_{b-}\gamma(t) \times (D^{1,\alpha}_{b-}\gamma(t) \times D^{\alpha}_{b-}\gamma(t))}{||D^{\alpha}_{b-}\gamma(t) \times (D^{1,\alpha}_{b-}\gamma(t) \times D^{\alpha}_{b-}\gamma(t))||}\right)$$

and similarly, given that $\mathbf{u}_3(t) = \mathbf{u}_1(t) \times \mathbf{u}_2(t)$, a few calculations reveal that

$$\mathbf{u}_3(t) = \frac{D^{\alpha}_{a+}\gamma(t) \times D^{1,\alpha}_{a+}\gamma(t)}{||D^{\alpha}_{a+}\gamma(t) \times D^{1,\alpha}_{a+}\gamma(t)||}$$

$$\left(\text{Respectively, } \mathbf{u}_3(t) = \frac{D^{\alpha}_{b-}\gamma(t) \times D^{1,\alpha}_{b-}\gamma(t)}{||D^{\alpha}_{b-}\gamma(t) \times D^{1,\alpha}_{b-}\gamma(t)||}\right).$$

Given that the left (resp. right) $\boldsymbol{\alpha}$-Frenet frame varies with the curve, it is natural to wonder how does this frame changes as the parameter *advances*. The following result provides an answer.

Theorem 1.8 *Let $\gamma \in \mathcal{F}_{a+}(\boldsymbol{\alpha}; (a, b), \mathbb{R}^n)$ (resp. $\gamma \in \mathcal{F}_{b-}(\boldsymbol{\alpha}; (a, b), \mathbb{R}^n)$). There exist $n - 1$ functions $\kappa^{\alpha}_{i,a+}$ (resp. $\kappa^{\alpha}_{i,b-}$) for $i = 1, \ldots, n - 1$ of class $C^{2p-n-i}((a, b))$, with $\kappa^{\alpha}_{i,a+}(t) > 0$ (resp. $\kappa^{\alpha}_{i,b-}(t) > 0$) for $1 \le i \le n - 2$, such that the left (resp. right) $\boldsymbol{\alpha}$-Frenet frame satisfies the differential equations*

$$\begin{pmatrix} \mathbf{u}_1'(t) \\ \vdots \\ \mathbf{u}_n'(t) \end{pmatrix} = ||D^{\alpha}_{a+}\gamma(t)|| \begin{pmatrix} 0 & \kappa^{\alpha}_{1,a+} & 0 & \cdots & 0 & 0 \\ -\kappa^{\alpha}_{1,a+} & 0 & \kappa^{\alpha}_{2,a+} & \cdots & 0 & 0 \\ 0 & -\kappa^{\alpha}_{2,a+} & 0 & \cdots & 0 & 0 \\ \vdots & \vdots & \vdots & \ddots & \vdots & \vdots \\ 0 & 0 & 0 & \cdots & 0 & \kappa^{\alpha}_{n-1,a+} \\ 0 & 0 & 0 & \cdots & -\kappa^{\alpha}_{n-1,a+} & 0 \end{pmatrix} \begin{pmatrix} \mathbf{u}_1(t) \\ \vdots \\ \mathbf{u}_n(t) \end{pmatrix}. \tag{1.29}$$

Respectively, for the right case,

$$\begin{pmatrix} \mathbf{u}_1'(t) \\ \vdots \\ \mathbf{u}_n'(t) \end{pmatrix} = ||D_{b-}^{\alpha}\gamma(t)|| \begin{pmatrix} 0 & \kappa_{1,b-}^{\alpha} & 0 & \cdots & 0 & 0 \\ -\kappa_{1,b-}^{\alpha} & 0 & \kappa_{2,b-}^{\alpha} & \cdots & 0 & 0 \\ 0 & -\kappa_{2,b-}^{\alpha} & 0 & \cdots & 0 & 0 \\ \vdots & \vdots & \vdots & \ddots & \vdots & \vdots \\ 0 & 0 & 0 & \cdots & 0 & \kappa_{n-1,b-}^{\alpha} \\ 0 & 0 & 0 & \cdots & -\kappa_{n-1,b-}^{\alpha} & 0 \end{pmatrix} \begin{pmatrix} \mathbf{u}_1(t) \\ \vdots \\ \mathbf{u}_n(t) \end{pmatrix}. \tag{1.30}$$

Proof As usual, we will prove the left case and leave the identical details of the right case to the reader.

The left $\boldsymbol{\alpha}$-Frenet frame fulfills the orthonormality relations

$$\langle \mathbf{u}_i(t), \mathbf{u}_j(t) \rangle = \delta_{ij}.$$

Hence, if we differentiate the previous relations, it follows that

$$\langle \mathbf{u}_i(t), \mathbf{u}_i'(t) \rangle = 0 \ \text{ and } \ \langle \mathbf{u}_i(t), \mathbf{u}_j'(t) \rangle = -\langle \mathbf{u}_i'(t), \mathbf{u}_j(t) \rangle \tag{1.31}$$

for $i \neq j$. In addition, for each $i = 1, \ldots, n-1$, we have that $\mathbf{u}_i(t) \in \text{span}\{\mathbf{u}_1(t), \ldots, \mathbf{u}_i(t)\} = \text{span}\{D_{a+}^{\alpha}\gamma(t), \ldots, D_{a+}^{i-1,\alpha}\gamma(t)\}$. Consequently, $\mathbf{u}_i'(t) \in \text{span}\{D_{a+}^{\alpha}\gamma(t), \ldots, D_{a+}^{i,\alpha}\gamma(t)\} = \text{span}\{\mathbf{u}_1(t), \ldots, \mathbf{u}_{i+1}(t)\}$. Therefore, by orthogonality,

$$\langle \mathbf{u}_i'(t), \mathbf{u}_j(t) \rangle = 0, \ j = i+2, \ldots, n.$$

Hence, by the relations (1.31), the only inner products of the derivatives $\mathbf{u}_i'(t)$ with the vectors $\mathbf{u}_j(t)$ that survive are $\langle \mathbf{u}_i'(t), \mathbf{u}_{i+1}(t) \rangle$ for $i = 1, \ldots, n-2$. Hence, let us define

$$\kappa_{i,a+}^{\alpha}(t) := \frac{\langle \mathbf{u}_i'(t), \mathbf{u}_{i+1}(t) \rangle}{||D_{a+}^{\alpha}\gamma(t)||}. \tag{1.32}$$

By Remark 1.11, each function $\kappa_{i,a+}^{\alpha}(t)$ belongs to the class $C^{2p-n-i}((a,b))$. Additionally, by orthogonality, the sign of $\langle \mathbf{u}_i'(t), \mathbf{u}_{i+1}(t) \rangle$ is the same as the sign of $\langle D_{a+}^{\alpha}\gamma(t), \mathbf{u}_{i+1}(t) \rangle$ because $\mathbf{u}_i'(t) \in \text{span}\{D_{a+}^{\alpha}\gamma(t), \ldots, D_{a+}^{i,\alpha}\gamma(t)\}$ as we already knew and by Proposition 1.5, it follows that $\kappa_{i,a+}^{\alpha}(t) > 0$ for $i = 1, \ldots, n-1$. □

From now on, the Eqs. (1.29) and (1.30) will be known as **left and right α-Frenet-Serret equations**, respectively, and for each $i = 1, \ldots, n-1$, the functions $\kappa_{i,a+}^{\alpha}$ and $\kappa_{i,b-}^{\alpha}$ will be denominated as ***i*-th left and right α-Frenet curvatures**, respectively.

Remark 1.12 The presence of the factors $||D_{a+}^{\alpha}\gamma(t)||$ and $||D_{b-}^{\alpha}\gamma(t)||$ in equations (1.29) and (1.30) is just a convention that serves to recover the classical curvatures when the differentiation orders become 1 ($\boldsymbol{\alpha} = \mathbf{1}$).

As a consequence of Theorem 1.8 and Proposition 1.4, the following result is immediate.

Corollary 1.8 *For $\gamma \in \mathcal{F}_{a+}(\boldsymbol{\alpha}; (a, b), \mathbb{R}^n)$ (resp. $\gamma \in \mathcal{F}_{b-}(\boldsymbol{\alpha}; (a, b), \mathbb{R}^n)$) such that γ is parametrized by left $\boldsymbol{\alpha}$-arc length (resp. right $\boldsymbol{\alpha}$-arc length), the left (resp. right) $\boldsymbol{\alpha}$-Frenet frame satisfies the differential equations*

$$\begin{pmatrix} \mathbf{u}_1'(t) \\ \vdots \\ \mathbf{u}_n'(t) \end{pmatrix} = \begin{pmatrix} 0 & \kappa_{1,a+}^{\boldsymbol{\alpha}} & 0 & \cdots & 0 & 0 \\ -\kappa_{1,a+}^{\boldsymbol{\alpha}} & 0 & \kappa_{2,a+}^{\boldsymbol{\alpha}} & \cdots & 0 & 0 \\ 0 & -\kappa_{2,a+}^{\boldsymbol{\alpha}} & 0 & \cdots & 0 & 0 \\ \vdots & \vdots & \vdots & \ddots & \vdots & \vdots \\ 0 & 0 & 0 & \cdots & 0 & \kappa_{n-1,a+}^{\boldsymbol{\alpha}} \\ 0 & 0 & 0 & \cdots & -\kappa_{n-1,a+}^{\boldsymbol{\alpha}} & 0 \end{pmatrix} \begin{pmatrix} \mathbf{u}_1(t) \\ \vdots \\ \mathbf{u}_n(t) \end{pmatrix}. \tag{1.33}$$

Respectively,

$$\begin{pmatrix} \mathbf{u}_1'(t) \\ \vdots \\ \mathbf{u}_n'(t) \end{pmatrix} = \begin{pmatrix} 0 & \kappa_{1,b-}^{\boldsymbol{\alpha}} & 0 & \cdots & 0 & 0 \\ -\kappa_{1,b-}^{\boldsymbol{\alpha}} & 0 & \kappa_{2,b-}^{\boldsymbol{\alpha}} & \cdots & 0 & 0 \\ 0 & -\kappa_{2,b-}^{\boldsymbol{\alpha}} & 0 & \cdots & 0 & 0 \\ \vdots & \vdots & \vdots & \ddots & \vdots & \vdots \\ 0 & 0 & 0 & \cdots & 0 & \kappa_{n-1,b-}^{\boldsymbol{\alpha}} \\ 0 & 0 & 0 & \cdots & -\kappa_{n-1,b-}^{\boldsymbol{\alpha}} & 0 \end{pmatrix} \begin{pmatrix} \mathbf{u}_1(t) \\ \vdots \\ \mathbf{u}_n(t) \end{pmatrix}. \tag{1.34}$$

Example 1.11 For $\boldsymbol{\alpha} \in (0, 1]^n$, let us consider $\gamma \in D_{a+}^{\mathbf{1}-\boldsymbol{\alpha}}(\mathcal{F}(1; (a, b), \mathbb{R}^n))$, whence $\gamma(t) = D_{a+}^{\mathbf{1}-\boldsymbol{\alpha}}\tilde{\gamma}(t)$ and $D_{a+}^{\boldsymbol{\alpha}}\gamma(t) = \tilde{\gamma}'(t)$; then it follows that the left $\boldsymbol{\alpha}$-Frenet-Serret equations and the left $\boldsymbol{\alpha}$-Frenet curvatures of γ will coincide with the classical Frenet-Serret equations and Frenet curvatures of $\tilde{\gamma}$, respectively, by the construction shown in the proof of Theorem 1.8 (c.f Lemma 4.1, p. 70 from [18]).

Example 1.12 The planar and spatial cases can be studied in detail.

If $n = 2$, only one fractional curvature is to be deduced. From Example 1.10, we find that

$$\mathbf{u}_1'(t) = \frac{||D_{a+}^{\boldsymbol{\alpha}}\gamma(t)||^2 D_{a+}^{1,\boldsymbol{\alpha}}\gamma(t) - \langle D_{a+}^{\boldsymbol{\alpha}}\gamma(t), D_{a+}^{1,\boldsymbol{\alpha}}\gamma(t)\rangle D_{a+}^{\boldsymbol{\alpha}}\gamma(t)}{||D_{a+}^{\boldsymbol{\alpha}}\gamma(t)||^3},$$

whence

$$\begin{aligned} \langle \mathbf{u}_1'(t), \mathbf{u}_2(t)\rangle &= \frac{\langle \mathbf{u}_2(t), D_{a+}^{1,\boldsymbol{\alpha}}\gamma(t)\rangle}{||D_{a+}^{\boldsymbol{\alpha}}\gamma(t)||} \\ &= \frac{\det(D_{a+}^{\boldsymbol{\alpha}}\gamma(t), D_{a+}^{1,\boldsymbol{\alpha}}\gamma(t))}{||D_{a+}^{\boldsymbol{\alpha}}\gamma(t)||^2}. \end{aligned}$$

Therefore,

$$\kappa_{1,a+}^{\alpha}(t) = \frac{\det(D_{a+}^{\alpha}\gamma(t), D_{a+}^{1,\alpha}\gamma(t))}{||D_{a+}^{\alpha}\gamma(t)||^3}.$$

Similarly,

$$\kappa_{1,b-}^{\alpha}(t) = \frac{\det(D_{b-}^{\alpha}\gamma(t), D_{b-}^{1,\alpha}\gamma(t))}{||D_{b-}^{\alpha}\gamma(t)||^3}.$$

On the other hand, if $n = 3$, we have two left (resp. right) $\boldsymbol{\alpha}$-Frenet curvatures, which can be expressed in terms of the parametrization γ. For $\kappa_{1,a+}^{\alpha}$ and $\kappa_{1,b-}^{\alpha}$, it is useful to deduce that

$$\begin{aligned}
\mathbf{u}_1'(t) &= \frac{D_{a+}^{\alpha}\gamma(t) \times (D_{a+}^{1,\alpha}\gamma(t) \times D_{a+}^{\alpha}\gamma(t))}{||D_{a+}^{\alpha}\gamma(t)||^3} \\
&= \frac{||D_{a+}^{\alpha}\gamma(t) \times (D_{a+}^{1,\alpha}\gamma(t) \times D_{a+}^{\alpha}\gamma(t))||}{||D_{a+}^{\alpha}\gamma(t)||^3}\mathbf{u}_2(t) \\
&= \frac{||D_{a+}^{1,\alpha}\gamma(t) \times D_{a+}^{\alpha}\gamma(t)||}{||D_{a+}^{\alpha}\gamma(t)||^2}\mathbf{u}_2(t) \\
&= ||D_{a+}^{\alpha}\gamma(t)||\kappa_{1,a+}^{\alpha}(t)\mathbf{u}_2(t),
\end{aligned}$$

where

$$\kappa_{1,a+}^{\alpha} = \frac{||D_{a+}^{1,\alpha}\gamma(t) \times D_{a+}^{\alpha}\gamma(t)||}{||D_{a+}^{\alpha}\gamma(t)||^3}.$$

Respectively, it can be deduced analogously that

$$\kappa_{1,b-}^{\alpha} = \frac{||D_{b-}^{1,\alpha}\gamma(t) \times D_{b-}^{\alpha}\gamma(t)||}{||D_{b-}^{\alpha}\gamma(t)||^3}.$$

Finally, for $\kappa_{2,a+}^{\alpha}$ and $\kappa_{2,b-}^{\alpha}$, after some long calculations we can find that

$$\begin{aligned}
\mathbf{u}_3'(t) &= -\frac{||D_{a+}^{\alpha}\gamma(t) \times (D_{a+}^{1,\alpha}\gamma(t) \times D_{a+}^{\alpha}\gamma(t))||}{||D_{a+}^{\alpha}\gamma(t) \times D_{a+}^{1,\alpha}\gamma(t)||^3}\langle D_{a+}^{\alpha}\gamma(t), D_{a+}^{1,\alpha}\gamma(t) \times D_{a+}^{2,\alpha}\gamma(t)\rangle\mathbf{u}_2(t) \\
&= -\frac{||D_{a+}^{\alpha}\gamma(t)||}{||D_{a+}^{\alpha}\gamma(t) \times D_{a+}^{1,\alpha}\gamma(t)||^2}\det(D_{a+}^{\alpha}\gamma(t), D_{a+}^{1,\alpha}\gamma(t), D_{a+}^{2,\alpha}\gamma(t))\mathbf{u}_2(t),
\end{aligned}$$

whence

$$\kappa_{2,a+}^{\alpha} = \frac{\det(D_{a+}^{\alpha}\gamma(t), D_{a+}^{1,\alpha}\gamma(t), D_{a+}^{2,\alpha}\gamma(t))}{||D_{a+}^{\alpha}\gamma(t) \times D_{a+}^{1,\alpha}\gamma(t)||^2}.$$

In a similar vein,

$$\kappa^{\boldsymbol{\alpha}}_{2,b-} = \frac{\det(D^{\boldsymbol{\alpha}}_{b-}\gamma(t), D^{1,\boldsymbol{\alpha}}_{b-}\gamma(t), D^{2,\boldsymbol{\alpha}}_{b-}\gamma(t))}{||D^{\boldsymbol{\alpha}}_{b-}\gamma(t) \times D^{1,\boldsymbol{\alpha}}_{b-}\gamma(t)||^2}.$$

Example 1.13 Let us consider the helix given by the parametrization $\gamma : [0, 2\pi] \to \mathbb{R}^3, \gamma(t) = (\cos t, \sin t, t)$. From the examples and results given in Sect. 1.1, it is easy to see that for any $t \in (0, 2\pi]$ and $\boldsymbol{\alpha} \in (0, 1]^3$,

$$D^{\boldsymbol{\alpha}}_{0+}\gamma(t) = \left(t^{-\alpha_1}E_{2,1-\alpha_1}(-t^2) - \frac{t^{\alpha_1}}{\Gamma(1-\alpha_1)}, t^{1-\alpha_2}E_{2,2-\alpha_2}(-t^2), \frac{t^{1-\alpha_3}}{\Gamma(2-\alpha_3)}\right),$$
$$D^{1,\boldsymbol{\alpha}}_{0+}\gamma(t) = \left(t^{-\alpha_1-1}E_{2,-\alpha_1}(-t^2) - \frac{t^{\alpha_1-1}}{\Gamma(-\alpha_1)}, t^{-\alpha_2}E_{2,1-\alpha_2}(-t^2), \frac{t^{-\alpha_3}}{\Gamma(1-\alpha_3)}\right) \text{ and}$$
$$D^{2,\boldsymbol{\alpha}}_{0+}\gamma(t) = \left(t^{-\alpha_1-2}E_{2,-1-\alpha_1}(-t^2) - \frac{t^{\alpha_1-2}}{\Gamma(-\alpha_1-1)}, t^{-\alpha_2-1}E_{2,-\alpha_2}(-t^2), \frac{t^{-\alpha_3-1}}{\Gamma(-\alpha_3)}\right).$$

In Fig. 1.4, we plotted some left $\boldsymbol{\alpha}$-Frenet frames and left $\boldsymbol{\alpha}$-Frenet curvatures for this curve, where we have used the previous fractional vectors. Moreover, we chose convenient order vectors $\boldsymbol{\alpha}$ to exhibit the limiting behavior of these moving frames and the corresponding curvatures as the orders α_i get closer to 1, which corresponds to the classical case of constant 1-Frenet curvatures equal to $1/2$ for the helix.

Remark 1.13 The reader may expect that the last $\boldsymbol{\alpha}$-Frenet curvature $\kappa^{\boldsymbol{\alpha}}_{n-1,a+}(t)$ (resp. $\kappa^{\boldsymbol{\alpha}}_{n-1,b-}(t)$) occupies a privileged place, just as the torsion does when $\boldsymbol{\alpha} = \mathbf{1}$, where zero torsion is equivalent to the curve belonging to a hyperplane in $\mathbb{R}^n$. However, this is not necessarily the case, as the following examples showcase.

Example 1.14 Consider $\gamma : [0, \infty) \to \mathbb{R}^2$ defined as

$$\gamma(t) = \left(\frac{t^{0.6}}{\Gamma(1.6)}, \frac{t^{1.8}}{\Gamma(2.8)}\right).$$

If $\boldsymbol{\alpha} := (0.6, 1.8)$, then $D^{\boldsymbol{\alpha}}_{0+}\gamma(t) = (1, 1)$ and consequently, $D^{1,\boldsymbol{\alpha}}_{0+}\gamma(t) = (0, 0)$. Therefore, $\kappa^{\boldsymbol{\alpha}}_{1,0+}(t) = 0$ by Example 1.12, but γ does not belong to any straight line (Fig. 1.5).

Example 1.15 Now let us consider $\boldsymbol{\alpha} := (\frac{1}{2}, \frac{3}{2}, \frac{1}{2})$ and $\gamma : [0, 2\pi] \to \mathbb{R}^3$ given as

$$\gamma(t) = (t^{1/2}E_{2,3/2}(-t^2), t^{5/2}E_{2,7/2}(-t^2), 1 - t^{1/2}E_{2,3/2}(-t^2) - t^{5/2}E_{2,7/2}(-t^2)).$$

It follows that

$$D^{\boldsymbol{\alpha}}_{0+}\gamma(t) = (\cos t, \sin t, -1),$$

$$D^{1,\boldsymbol{\alpha}}_{0+}\gamma(t) = (-\sin t, \cos t, 0) \text{ and}$$

$$D^{2,\boldsymbol{\alpha}}_{0+}\gamma(t) = (-\cos t, -\sin t, 0).$$

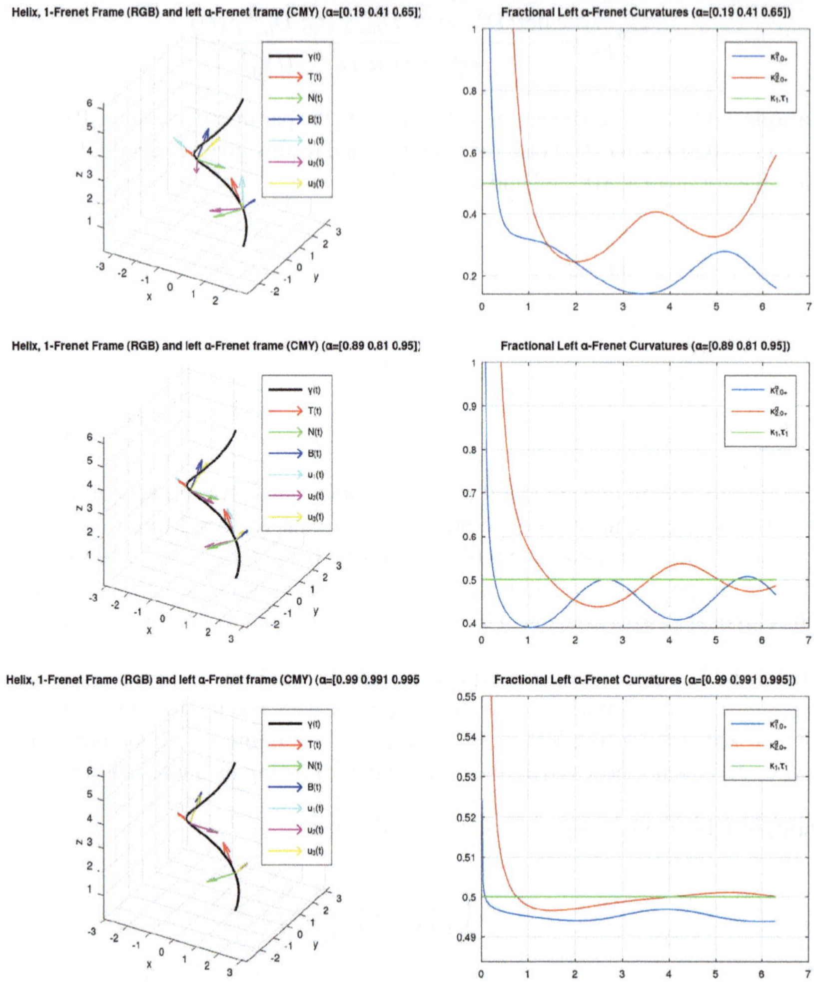

Fig. 1.4 Comparison between the classical and fractional Frenet frames and Frenet curvatures for the helix of Example 1.13, for distinct choices of the vector $\boldsymbol{\alpha} \in (0, 1]^3$

By Example 1.12, we find that $\kappa^{\boldsymbol{\alpha}}_{2,0+}(t) = -\frac{1}{2}\cos t$. Therefore, the last left $\boldsymbol{\alpha}$-Frenet curvature of γ does not vanish on the entire interval $[0, 2\pi]$, but γ still belongs to the plane described by the equation $x + y + z = 1$ (Fig. 1.5).

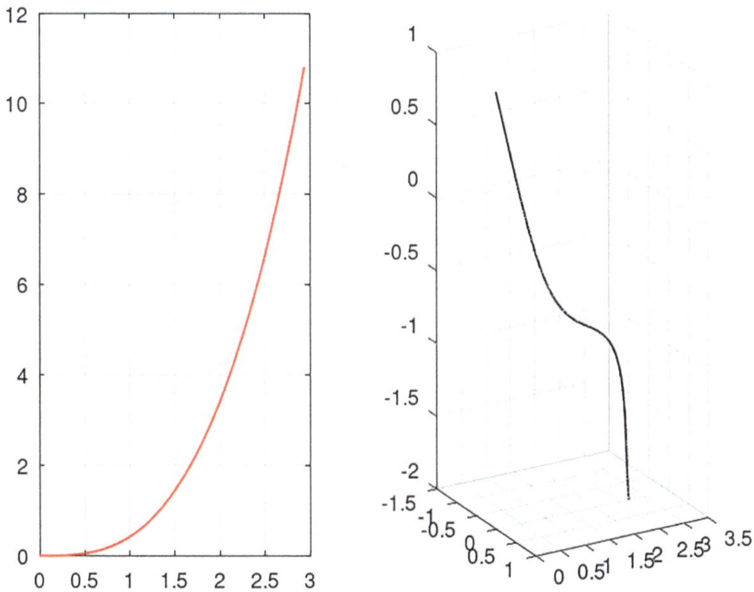

Fig. 1.5 On the left, the plane curve from Example 1.14 and on the right, the curve from Example 1.15

1.4 A Fundamental Theorem of Curves of Order α

If we take a look at the systems (1.29) and (1.30), given the respective $\boldsymbol{\alpha}$-Frenet curvatures, such systems could be integrated upon initial conditions and we should be able to recover respective $\boldsymbol{\alpha}$-Frenet curves. This process constitutes the inverse reasoning of Theorem 1.8 and it is the main result of this work.

Theorem 1.9 *Let $a \in (c, d)$ (resp. $b \in (c, d)$) and consider $n - 1$ functions $\kappa^{\alpha}_{i,a+}$ (resp. $\kappa^{\alpha}_{i,b-}$) in $C^{2p-n-i}((c, d))$ for each $i = 1, \dots, n - 1$, where $\kappa^{\alpha}_{i,a+} > 0$ (resp. $\kappa^{\alpha}_{i,b-} > 0$) for $i = 1, \dots, n - 2$. In addition, let us fix $\mathbf{x}^{(0)}_0, \dots, \mathbf{x}^{(0)}_{m-1} \in \mathbb{R}^n$ and n vectors $\mathbf{u}^{(0)}_1, \dots, \mathbf{u}^{(0)}_n$. There exists a unique (up to rigid transformations of $\mathbb{R}^n$) left (resp. right) $\boldsymbol{\alpha}$-regular curve $\gamma \in C^{2p-n}((a, d), \mathbb{R}^n)$ (resp. $\gamma \in C^{2p-n}((c, b))$), parametrized by left (resp. right) $\boldsymbol{\alpha}$-arc length, such that*

1. *$\gamma(a) = \mathbf{x}^{(0)}_0$ (resp. $\gamma(b) = \mathbf{x}^{(0)}_0$);*
2. *$\mathbf{u}^{(0)}_1, \dots, \mathbf{u}^{(0)}_n$ is the left (resp. right) $\boldsymbol{\alpha}$-Frenet frame of γ at $\mathbf{x}^{(0)}_0$;*
3. *the given functions $\kappa^{\alpha}_{i,a+}$ (resp. $\kappa^{\alpha}_{i,b-}$) are the left (resp. right) $\boldsymbol{\alpha}$-Frenet curvatures of γ.*

Proof

Part 1. Given the $n - 1$ functions $\kappa^{\alpha}_{i,a+}$ (resp. $\kappa^{\alpha}_{i,b-}$), let us consider the linear system of ordinary differential equations given in (1.33) (resp. in (1.34)), which consists of n^2 differential equations, subject to initial (resp. terminal) conditions $\mathbf{u}_i(a) = \mathbf{u}_i^{(0)}$ (resp. $\mathbf{u}_i(b) = \mathbf{u}_i^{(0)}$) for $i = 1, \ldots, n-1$. Due to the existence and uniqueness theorem for ordinary differential equations (c.f. Theorem 3.1, p. 553 from [19]), (1.33) (resp. (1.34)) possesses a unique solution[2] $\big(\mathbf{u}_1(t) \cdots \mathbf{u}_n(t)\big)$ defined on the whole open interval (a, d) (resp. (c, b)). Moreover, given that the functions $\kappa^{\boldsymbol{\alpha}}_{i,a+}$ (resp. $\kappa^{\boldsymbol{\alpha}}_{i,b-}$) belong to the class $C^{2p-n-i}((a, d))$ (resp. $C^{2p-n-i}((c, b))$), it follows that $\big(\mathbf{u}_1 \cdots \mathbf{u}_n\big) \in C^{2p-n}((a, d), \mathbb{R}^{n\times n})$ (resp. in $C^{2p-n}((c, b), \mathbb{R}^{n\times n})$).

Part 2. Now let us prove that if the solution $\big(\mathbf{u}_1(t) \cdots \mathbf{u}_n(t)\big)$ begins (resp. *ends*) orthonormal and positively oriented, it will remain (resp. it remained) like that for the rest of the interval (a, d) (resp. (c, b)). For such purpose, let us consider the Gram matrix $G(t)$ associated to the vectors $\{\mathbf{u}_1(t), \ldots, \mathbf{u}_n(t)\}$, given by the matrix function whose entries are the inner products $\langle \mathbf{u}_i(t), \mathbf{u}_j(t)\rangle$ for $i, j = 1, \ldots, n$. Moreover, by the orthonormality and orientability hypotheses, let us suppose that $G(a) = \mathrm{Id}$ (resp. $G(b) = \mathrm{Id}$). In addition, let us observe that if we define $F(t) := \big(\mathbf{u}_1(t) \cdots \mathbf{u}_n(t)\big)$, it follows that $G(t) := F^T(t)F(t)$. Hence, if we 1-differentiate the matrix $G(t)$, we will find that

$$\begin{aligned}
G'(t) &= F'(t)^T F(t) + F(t)^T F'(t) \\
&= (F'(t))^T F(t) + F(t)^T F'(t) \\
&= (F(t)K(t))^T F(t) + F(t)^T F(t)K(t) \\
&= K(t)^T F(t)^T F(t) + F(t)^T F(t)K(t) \\
&= K(t)^T G(t) + G(t)K(t),
\end{aligned}$$

where the matrix function $K(t)$ is the matrix of left (resp. right) $\boldsymbol{\alpha}$-Frenet curvatures from system (1.29) (resp. (1.30)). Now notice that the identity matrix is a solution of the system $G'(t) = K(t)^T G(t) + G(t)K(t)$:

$$\begin{aligned}
0 &= (\mathrm{Id})' \\
&= K(t)^T + K(t) \\
&= 0,
\end{aligned}$$

where the last equality follows by the antisymmetry of $K(t)$. By existence and uniqueness of the solution of the linear system $G'(t) = K^T(t)G(t) + G(t)K(t)$ with initial (resp. terminal) condition $G(a) = \mathrm{Id}$ (resp. $G(b) = \mathrm{Id}$), we can conclude that $G(t) = \mathrm{Id}$ for any $t \in (a, d)$ (resp. $t \in (c, b)$) and the solution $F(t)$ is always orthonormal and linearly independent in

[2] In the case of the system (1.34) with terminal values $\mathbf{u}_i(b) = \mathbf{u}_i^{(0)}$, this problem can be easily transformed into an equivalent initial value problem with the reflection operator Q from Proposition 1.1 applied to the unknown vector functions $\mathbf{u}_i(t)$, whose solution will be the desired solution for our terminal value problem.

consequence. Finally, the positive orientation of the frame $\{\mathbf{u}_1(t), \ldots, \mathbf{u}_n(t)\}$ is a consequence of the continuity of the determinant function.[3]

Part 3. Now that we have recovered a candidate for the left (resp. right) $\boldsymbol{\alpha}$-Frenet frame $\{\mathbf{u}_1(t), \ldots, \mathbf{u}_n(t)\}$, let us reconstruct a curve from it. For such purpose, let us consider the system of left (resp. right) fractional differential equations

$$D^{\boldsymbol{\alpha}}_{a+}\gamma(t) = \mathbf{u}_1(t) \text{ (resp. } D^{\boldsymbol{\alpha}}_{b-}\gamma(t) = \mathbf{u}_1(t)), \tag{1.35}$$

subject to the initial (resp. *terminal*) conditions

$$D^{\mathbf{k}}_{a+}\gamma(a) = \mathbf{x}^{(0)}_k \text{ (resp. } D^{\mathbf{k}}_{b-}\gamma(b) = \mathbf{x}^{(0)}_k) \tag{1.36}$$

for $k = 0, 1, \ldots, m - 1$.

The system (1.35), explicitly, looks like

$$\begin{aligned} D^{\alpha_1}_{a+}\gamma_1(t) &= u_{11}(t) \text{ (resp. } D^{\alpha_1}_{b-}\gamma_1(t) = u_{1,1}(t)) \\ &\vdots \\ D^{\alpha_n}_{a+}\gamma_n(t) &= u_{1n}(t) \text{ (resp. } D^{\alpha_n}_{b-}\gamma_n(t) = u_{1,n}(t)) \end{aligned}$$

Fortunately, system (1.35) can be solved explicitly by means of Theorem 1.3: for each $i = 1, \ldots, n$, it follows that

$$\begin{aligned} I^{\alpha_i}_{a+}D^{\alpha_i}_{a+}\gamma_i(t) &= \gamma_i(t) - \sum_{k=0}^{m_i-1} \frac{\gamma_i^{(k)}(a)}{k!}(t-a)^k \\ &= I^{\alpha_i}_{a+}u_{1,i}(t), \end{aligned}$$

then

$$\gamma_i(t) = I^{\alpha_i}_{a+}u_{1,i}(t) + \sum_{k=0}^{m_i-1} \frac{\gamma_i^{(k)}(a)}{k!}(t-a)^k$$

(Respectively,

$$\begin{aligned} I^{\alpha_i}_{b-}D^{\alpha_i}_{b-}\gamma_i(t) &= \gamma_i(t) - \sum_{k=0}^{m_i-1} \frac{(-1)^k\gamma_i^{(k)}(b)}{k!}(b-t)^k \\ &= I^{\alpha_i}_{b-}u_{1,i}(t), \end{aligned}$$

[3] The justification for the terminal value problem $G' = K^T G + GK$ is the same as before: it can be transformed into an IVP via the reflection operator Q from Proposition 1.1.

thus,

$$\gamma_i(t) = I_{b-}^{\alpha_i} u_{1,i}(t) + \sum_{k=0}^{m_i-1} \frac{(-1)^k \gamma_i^{(k)}(b)}{k!}(b-t)^k).$$

By the given initial (resp. terminal) conditions (1.36), considering that

$$\gamma_i^{(k)}(a) = x_{k,i}^{(0)} \text{ resp. } \gamma_i^{(k)}(b) = (-1)^k x_{k,i}^{(0)}$$

for $k = 0, \ldots, m-1$ and $i = 1, \ldots, n$, it follows that

$$\gamma_i(t) = I_{a+}^{\alpha_i} u_{1,i}(t) + \sum_{k=0}^{m_i-1} \frac{x_{k,i}^{(0)}}{k!}(t-a)^k$$

$$(\text{Respectively, } \gamma_i(t) = I_{b-}^{\alpha_i} u_{1,i}(t) + \sum_{k=0}^{m_i-1} \frac{x_{k,i}^{(0)}}{k!}(b-t)^k).$$

The curve γ that we have obtained belongs to the class $C^{2p-n}((a,d), \mathbb{R}^n)$ (resp. $C^{2p-n}((c,b), \mathbb{R}^n)$) because the vector function $\mathbf{u}_1(t)$ does, it is parametrized by left (resp. right) $\boldsymbol{\alpha}$-arc length because $\mathbf{u}_1(t)$ is a unit vector (Proposition 1.4) and it satisfies condition 1 by construction. We are left to prove the remaining conditions and the uniqueness up to rigid transformations of $\mathbb{R}^n$.

Part 4. In order to prove conditions 2 and 3, let us differentiate the vectors that we obtained in the first step.

$$\begin{aligned}
\mathbf{u}_1'(t) &= D_{a+}^{1,\boldsymbol{\alpha}} \gamma(t) \\
&= \kappa_{1,a+}^{\boldsymbol{\alpha}}(t) \mathbf{u}_2(t). \\
\mathbf{u}_2'(t) &= D_{a+}^{2,\boldsymbol{\alpha}} \gamma(t) \\
&= \frac{d\kappa_{1,a+}^{\boldsymbol{\alpha}}(t)}{dt} \mathbf{u}_2(t) + \kappa_{1,a+}^{\boldsymbol{\alpha}}(t) \mathbf{u}_2'(t) \\
&= \frac{d\kappa_{1,a+}^{\boldsymbol{\alpha}}(t)}{dt} \mathbf{u}_2(t) - \kappa_{1,a+}^{\boldsymbol{\alpha}}(t)^2 \mathbf{u}_1(t) + \kappa_{1,a+}^{\boldsymbol{\alpha}}(t) \kappa_{2,a+}^{\boldsymbol{\alpha}}(t) \mathbf{u}_3(t).
\end{aligned}$$

Continuing with this process, we see that for $i = 0, \ldots, n-2$, $D_{a+}^{i,\boldsymbol{\alpha}} \gamma(t)$ is a linear combination of the vectors $\mathbf{u}_j(t)$ with $j = 1, \ldots, i+1$. As a consequence, the linear independence of the orthonormal set $\{\mathbf{u}_1(t), \ldots, \mathbf{u}_n(t)\}$ is inherited by the set $\{D_{a+}^{\boldsymbol{\alpha}} \gamma(t), \ldots, D_{a+}^{n-2,\boldsymbol{\alpha}} \gamma(t)\}$, due to the positivity of the functions $\kappa_{i,a+}^{\boldsymbol{\alpha}}$ (resp. $\kappa_{i,b-}^{\boldsymbol{\alpha}}$). Therefore, γ is left $\boldsymbol{\alpha}$-Frenet. An identical line of reasoning follows for the right case and correspondingly, the set $\{D_{b-}^{\boldsymbol{\alpha}} \gamma(t), \ldots, D_{b-}^{n-2,\boldsymbol{\alpha}} \gamma(t)\}$ will be linearly independent and γ will be right $\boldsymbol{\alpha}$-Frenet. In addition, given that the set $\{\mathbf{u}_1(t), \ldots, \mathbf{u}_{n-1}(t)\}$ spans the same subspace as

$\{D^{\alpha}_{a+}\gamma(t), \ldots, D^{n-2,\alpha}_{a+}\gamma(t)\}$ (resp. $\{D^{\alpha}_{b-}\gamma(t), \ldots, D^{n-2,\alpha}_{b-}\gamma(t)\}$), the positivity of the $\boldsymbol{\alpha}$-Frenet curvatures and the positive orientability of the $\{\mathbf{u}_1(t), \ldots, \mathbf{u}_{n-1}(t)\}$ imply that this is precisely the left (resp. right) $\boldsymbol{\alpha}$-Frenet frame of γ. As a consequence, conditions 2 and 3 are fulfilled.

Part 5. Finally, let us suppose that there exists another curve $\tilde{\gamma} \in C^{2p-n}((a, d), \mathbb{R}^n)$ (resp. in $C^{2p-n}((c, b), \mathbb{R}^n)$) with left $\boldsymbol{\alpha}$-Frenet (resp. right) curvatures $\tilde{\kappa}^{\alpha}_{i,a+} = \kappa^{\alpha}_{i,a+}$ (resp. $\tilde{\kappa}^{\alpha}_{i,b-} = \kappa^{\alpha}_{i,b-}$) and left (resp. right) $\boldsymbol{\alpha}$-Frenet frame $\{\mathbf{w}_1(t), \ldots, \mathbf{w}_n(t)\}$; this frame and $\{\mathbf{u}_1(t), \ldots, \mathbf{u}_n(t)\}$ are both positive oriented, so there exists some orientation-preserving rigid transformation $M(t) \in E(n)$ such that $\{\mathbf{w}_1(t), \ldots, \mathbf{w}_n(t)\}$ and $\{\mathbf{u}_1(t), \ldots, \mathbf{u}_n(t)\}$ coincide. Hence, the curve $M \circ \tilde{\gamma}$ possesses the same left (resp. right) $\boldsymbol{\alpha}$-Frenet frame and $\boldsymbol{\alpha}$-Frenet curvatures as γ over (a, d) (resp. (c, b)) and therefore, both curves satisfy (1.33) (resp. (1.34)). By uniqueness of the solution of the equations (1.35), $M \circ \tilde{\gamma} = \gamma$. □

As expected, explicit solutions derived from Theorem 1.9 are difficult to obtain, in general. However, simple but remarkable cases of this theorem stem when we consider **constant** left (resp. right) $\boldsymbol{\alpha}$-Frenet curvatures.

Example 1.16 In the plane ($n = 2$), let us consider $\kappa^{\alpha}_{1,0+}$ constant over $[0, 2\pi]$. Hence, the system (1.33) becomes easy to solve, given that it can be reduced to the simpler system

$$\mathbf{u}_1''(t) + \left(\kappa^{\alpha}_{1,a+}\right)^2 \mathbf{u}_1(t) = \mathbf{0}.$$

Therefore,

$$\mathbf{u}_1(t) = \mathbf{e}\cos(\kappa^{\alpha}_{1,a+}t) + \mathbf{f}\sin(\kappa^{\alpha}_{1,a+}t).$$

From the foregoing proof and Example 1.3, it follows that

$$\gamma(t) = \begin{pmatrix} e_1 t^{\alpha_1} E_{2,1+\alpha_1}(-(\kappa^{\alpha}_{1,a+}t)^2) + f_1 t^{\alpha_1+1} E_{2,2+\alpha_1}(-(\kappa^{\alpha}_{1,a+}t)^2) + \sum_{k=0}^{m_1-1} \frac{x^{(0)}_{k,1}}{k!} t^k \\ e_2 t^{\alpha_2} E_{2,1+\alpha_2}(-(\kappa^{\alpha}_{1,a+}t)^2) + f_2 t^{\alpha_2+1} E_{2,2+\alpha_2}(-(\kappa^{\alpha}_{1,a+}t)^2) + \sum_{k=0}^{m_2-1} \frac{x^{(0)}_{k,2}}{k!} t^k \end{pmatrix}.$$

A couple of special cases of the previous family of curves were plotted in Fig. 1.6.

For a spatial example ($n = 3$), now we consider $\kappa^{\alpha}_{1,2\pi-}$ and $\kappa^{\alpha}_{2,2\pi-}$ both constant and non-vanishing in $[0, 2\pi]$. In addition, let us set

$$\lambda^2 := (\kappa^{\alpha}_{1,2\pi-})^2 + \frac{(\kappa^{\alpha}_{2,2\pi-})^2}{\kappa^{\alpha}_{1,2\pi-}} > 0.$$

For the sake of simplicity, let us choose $\kappa^{\alpha}_{1,2\pi-}$ and $\kappa^{\alpha}_{2,2\pi-}$ such that $\lambda = 1$. Analogously to the previous example, the system (1.34) can be reduced to

$$\mathbf{u}_1'''(t) + \mathbf{u}_1'(t) = \mathbf{0}.$$

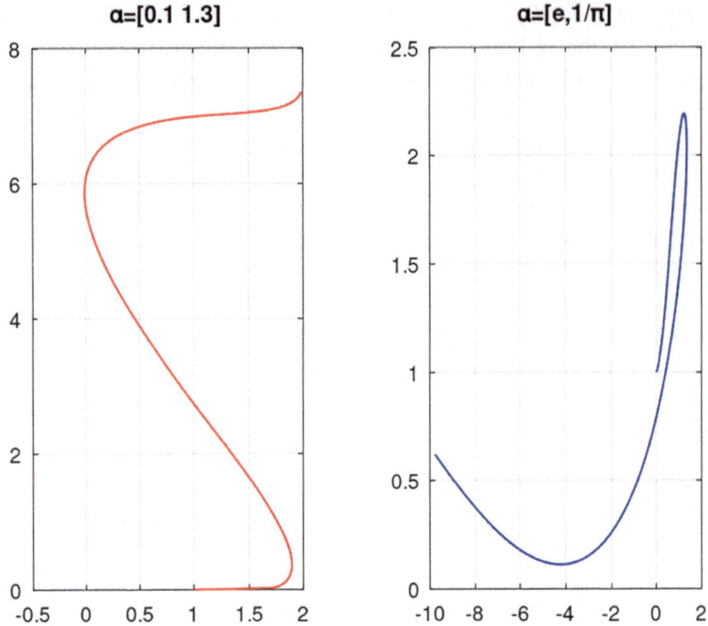

Fig. 1.6 A pair of plane curves with $\kappa^{\boldsymbol{\alpha}}_{1,0+} = 1$. On the left, the parameters are $\boldsymbol{\alpha} = (0.1, 1.3)$, $\mathbf{u}_1^{(0)} = (1, 0)$, $\mathbf{u}_2^{(0)} = (0, 1)$, $\mathbf{x}_0^{(0)} = (1, 0)$ and $\mathbf{x}_1^{(0)} = (0, 1)$. On the right, we chose $\boldsymbol{\alpha} = (e, 1/\pi)$, $\mathbf{u}_1^{(0)} = (1, 0)$, $\mathbf{u}_2^{(0)} = (0, 1)$, $\mathbf{x}_0^{(0)} = (0, 1)$, $\mathbf{x}_1^{(0)} = (1, 0)$ and $\mathbf{x}_2^{(0)} = (-1, -1)$

In a similar vein as the previous case,

$$\mathbf{u}_1'(t) = \mathbf{e}\cos(t) + \mathbf{f}\sin(t),$$

whence

$$\mathbf{u}_1(t) = \mathbf{g} + \mathbf{e}\sin(t) - \mathbf{f}\cos(t).$$

Finally, by means of Example 1.3 and the proof of Theorem 1.9, we conclude that

$$\gamma(t) = \begin{pmatrix} \frac{g_1(2\pi - t)^{\alpha_1}}{\Gamma(\alpha_1+1)} - e_1(2\pi - t)^{1+\alpha_1} E_{2,2+\alpha_1}(-(2\pi - t)^2) \\ \frac{g_2(2\pi - t)^{\alpha_2}}{\Gamma(\alpha_2+1)} - e_2(2\pi - t)^{1+\alpha_2} E_{2,2+\alpha_2}(-(2\pi - t)^2) \\ \frac{g_3(2\pi - t)^{\alpha_3}}{\Gamma(\alpha_3+1)} - e_3(2\pi - t)^{1+\alpha_3} E_{2,2+\alpha_3}(-(2\pi - t)^2) \end{pmatrix}$$
$$- \begin{pmatrix} f_1(2\pi - t)^{\alpha_2} E_{2,1+\alpha_2}(-(2\pi - t)^2) - \sum_{k=0}^{m_1-1} \frac{x_{k,1}^{(0)}}{k!}(2\pi - t)^k \\ f_2(2\pi - t)^{\alpha_2} E_{2,1+\alpha_2}(-(2\pi - t)^2) - \sum_{k=0}^{m_2-1} \frac{x_{k,2}^{(0)}}{k!}(2\pi - t)^k \\ f_3(2\pi - t)^{\alpha_3} E_{2,1+\alpha_3}(-(2\pi - t)^2) - \sum_{k=0}^{m_3-1} \frac{x_{k,3}^{(0)}}{k!}(2\pi - t)^k \end{pmatrix}.$$

A pair of particular cases have been plotted in Fig. 1.7.

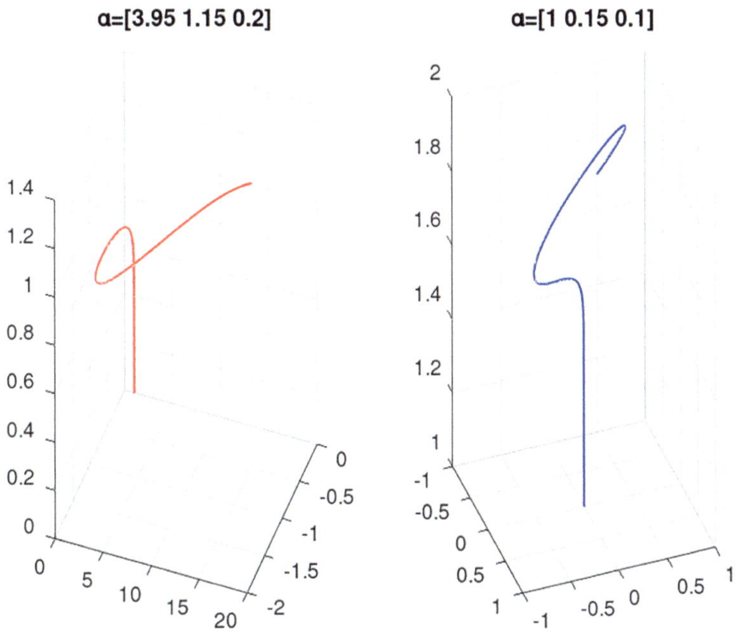

Fig. 1.7 A pair of space curves with $\kappa^{\alpha}_{1,2\pi-} = 1/2$ and $\kappa^{\alpha}_{2,2\pi-} = \sqrt{6}/4$. On the left, the parameters are $\boldsymbol{\alpha} = (3.95, 1.15, 0.2)$, $\mathbf{u}_1^{(0)} = (1, 0, 1)/\sqrt{2}$, $\mathbf{u}_2^{(0)} = (0, 1, 0)$, $\mathbf{u}_3^{(0)} = (-1, 0, 1)/\sqrt{2}$, $\mathbf{x}_0^{(0)} = (1, 0, 0)$ and $\mathbf{x}_1^{(0)} = \mathbf{x}_2^{(0)} = \mathbf{x}_3^{(0)} = (0, 0, 0)$. On the right, we chose $\boldsymbol{\alpha} = (1, 0.15, 0.1)$, the same initial right $\boldsymbol{\alpha}$-Frenet frame as before and $\mathbf{x}_1^{(0)} = (0, 0, 1)$

1.5 Conclusions

Throughout this research, we used basic notions and results from the theory of Riemann-Liouville integrals and Caputo derivatives to generalize classical concepts from the theory of curves in Euclidean spaces to a fractional setting, such as regularity, arc-length and Frenet curves. This way, we obtained the $\boldsymbol{\alpha}$-Frenet-Serret formulae, a new notion of $\boldsymbol{\alpha}$-curvatures and a fundamental theorem of order $\boldsymbol{\alpha}$ to recover curves from the given $\boldsymbol{\alpha}$-curvatures (up to rigid motions), for vectors $\boldsymbol{\alpha}$ of arbitrary real positive orders.

Acknowledgements The first author would like to express his acknowledgements to CONAHCYT (No. of CVU 1312206) for their support.

Competing Interests The authors do not declare any competing interest.

References

1. Euler L (1741) De constructione æquationum ope motus tractorii aliisque ad methodum tangentium inversam pertinentibus. Commentarii academiæ scientiarum Petropolitanæ, 66–85
2. Hoppe R (1862) Ueber die Darstellung der Curven durch Krümmung und Torsion. Walter de Gruyter, Berlin/New York
3. Frenet F (1852) Sur les courbes à double courbure. Journal de mathématiques pures et appliquées 17:437–447
4. Serret JA (1851) Sur quelques formules relatives à la théorie des courbes à double courbure. Journal de mathématiques pures et appliquées 16:193–207
5. Yajima T, Nagahama H (2010) Differential geometry of viscoelastic models with fractional-order derivatives. J Phys A: Mathem Theoret 43(38):385207
6. Yajima T, Yamasaki K (2012) Geometry of surfaces with Caputo fractional derivatives and applications to incompressible two-dimensional flows. J Phys A: Mathem Theoret 45(6):065201
7. Lazopoulos KA, Lazopoulos AK (2016) Fractional differential geometry of curves & surfaces. Progr Fract Differ Appl 2(3):169–186
8. Yajima T, Oiwa S, Yamasaki K (2018) Geometry of curves with fractional-order tangent vector and Frenet-Serret formulas. Fract Calculus Appl Anal 21(6):1493–1505
9. Aydın ME, BektasÌ§, M, ÖğremisÌ§, A, YokusÌ§ A (2021) Differential geometry of curves in Euclidean 3-space with fractional order. Int Electron J Geometry 14(1):132–144
10. ÖğrenmisÌ§ M (2022) Geometry of curves with fractional derivatives in Lorentz plane. J New Theory 38:88–98
11. Lopez FR, Rubio O (2023) A new fractional curvature of curves using the caputo's fractional derivative. Adv Math Mod Appl 8(2):157–175
12. Samko SG, Kilbas AA, Marichev OI (1993) Fractional integrals and derivatives. Theory and applications. Gordon & Breach Science Publishers
13. Diethelm K, Ford NJ (2010) The analysis of fractional differential equations. Lecture notes in mathematics. Springer
14. Kilbas AA, Srivastava HM, Trujillo JJ (2006) Theory and applications of fractional differential equations (Vol 204). Elsevier
15. Royden H, Fitzpatrick PM (2010) Real analysis. 4.a ed. China Machine Press
16. Apostol TM (1991) Calculus, vol 1. Wiley
17. Couto IT, Lymberopoulos A (2021) Introduction to Lorentz geometry: curves and surfaces. Chapman and Hall/CRC
18. O'Neil B (2006) Elementary differential geometry. Elsevier
19. Lang S (2013) Undergraduate analysis. Springer Science & Business Media

Chapter 2
From Self-similar Networks to Fractional Systems. A Formal Introduction

Jesús Enrique Escalante-Martínez, Adrian-Josue Guel-Cortez, Liliana Félix, and César Fernando Méndez-Barrios

2.1 Introduction

Networks are a ubiquitous option to describe complex systems as they easily help representing intricate interconnections, flow of information or mass [1]. Yet, finding model reductions to describe local or global aspects of networks into a more manageable system is a preferred practical approach [2]. Within the use of networks, it oversees their use in describing the transfer, flow, or transport of a quantity through their branches. For instance, electrical current, heat transfer, or the movement of mass through the network's pathways [3].

In the seminal work by Mayes and Sen [4], the authors introduced a model reduction approach for large-scale flow networks based on the network's self-similarity structure. Interestingly, the method results in a mathematical representation of the network in the form of a fractional-order operator. Later, Sen et al. [5] presented a generalization of the approach, applying it to both ladder and tree-like networks of linear lumped elements. These findings have led to applications in fields such as robotics [6], control theory [7], and cyber-physical systems [8].

Since this approach results in a workable model while also providing some physical interpretation for fractional-order operators, a well-discussed topic [9, 10], we

J. E. Escalante-Martínez
Facultad de Ingeniería Mecánica y Eléctrica, Universidad Veracruzana, Poza Rica, Veracruz, Mexico
e-mail: jeescalante@uv.mx

A.-J. Guel-Cortez
Brill Power, Oxford, United Kingdom

L. Félix · C. F. Méndez-Barrios (✉)
Facultad de Ingeniería, Universidad Autónoma de San Luis Potosí, San Luis Potosí, México
e-mail: fernando.barrios@uaslp.mx

L. Félix
e-mail: lfelixa_@uaslp.mx

J. J. Rosales García et al. (eds.), *Advances in Fractional Calculus*,
Modeling and Optimization in Science and Technologies 21,
https://doi.org/10.1007/978-3-031-83989-4_2

aim in this work to formally present the model reduction methodology while also delving deeper into the analytical properties of the networks and exploring new practical applications.

Specifically, we first detail and review the self-similarity conditions under which certain networks of linear lumped elements converge to fractional-order operators described using the Caputo and Riemann-Liouville definitions. Then, based on [11], we address the conditions under which any commensurable order can be obtained for a given model reduction. Finally, we study the dynamics of cardiac muscle using fractional-order operators. The proposed model is derived from [12, 13] and related citations, resulting in a model that qualitatively captures not only certain normal dynamics of cardiac muscle but also prevalent cardiovascular diseases. The latter is based on the pressure-volume curve characteristics associated with common cardiovascular conditions such as aortic regurgitation, mitral stenosis, and ischemic cardiomyopathy.

Notations. The following standard notations are used: the set of real numbers is denoted by $\mathbb{R}$. In particular, $\mathbb{R}_+ := \{x \in \mathbb{R} : x > 0\}$. Next, $\mathbb{Z}$ ($\mathbb{Z}_+, \mathbb{Z}_-$) represents the set of integers (strictly positive/negative). Similarly, the natural numbers are defined as $\mathbb{N} := \mathbb{Z}_+$.

2.2 Potential-Induced Flows in Self-similar Networks

This section introduces some definitions and results that will allow us to present formally the concept of self-similar networks with potential-driven flow, as well as, how this formalism can help to represent some complex systems.

Definition 2.1 (*Self-similar network*) Let $i, j, N \in \mathbb{N}$, such that $1 \leq i \leq N$, $1 \leq j \leq 2^N$. A self-similar network is a collection of branches denoted by (i, j) and branching points, such that in the generation N, there are a total of 2^N admisible paths, $2^{N+1} - 2$ branches and $2^{N+1} - 1$ branching points. See, for instance, Fig. 2.1.

Remark 2.1 It is worth mentioning, that in this note, by flow we refer to a flow of fluid, electric, energy, heat, etc.

Definition 2.2 (*Self-similar network with potential-driven flow (SSNPF)* Let $q := q_{in}$ denote the quantity flowing through the branches of a self-similar network, $\Delta P := P_{in} - P_{out}$ represent the potential difference driving the flow of q, and let L be a linear operator that relates q to ΔP. We will assume that the operator L is injective (i.e., $\ker(L) = \{0\}$) and surjective (i.e., the image of L must cover the entire space). Suppose the following conditions hold:

(i) The flow occurs along the branches;
(ii) The flow is driven by a potential difference;
(iii) The quantity $q_{i,j}$ is conserved at each branching point;
(iv) All branches have zero initial conditions.

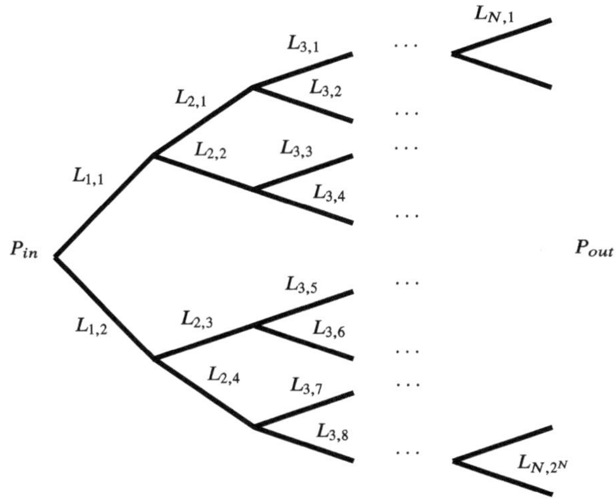

Fig. 2.1 Self-similar network with N-generation

Under these assumptions, the relationship for each branch (i, j) between $L_{i,j}$, $q_{i,j}$, and $\Delta P_{i,j}$ is described by the following transfer function:

$$L_{i,j}q_{i,j} = \Delta P_{i,j}, \tag{2.1}$$

where $q_{i,j}$ is the quantity transferred through branch (i, j), $\Delta P_{i,j}$ is the potential difference across branch (i, j), and $L_{i,j}$ is the operator relating $q_{i,j}$ to $\Delta P_{i,j}$ for branch (i, j).

Definition 2.3 (*Adjacent Branching Points*) Let $P_{j,k}$ and $P_{i,\ell}$ be two consecutive branching points. We define $P_{j,k}$ and $P_{i,\ell}$ as adjacent if the indices (i, ℓ) satisfy the following conditions: $i \equiv j - 1$ and $\ell \equiv \lceil \frac{k}{2} \rceil$. In this case, we denote adjacent branching points by

$$P_{j,k} \sim P_{j-1,\lceil \frac{k}{2} \rceil}, \quad 1 \leq j \leq N, \quad 1 \leq k \leq 2^N. \tag{2.2}$$

If $P_{j,k}$ is not adjacent to $P_{i,\ell}$, we write $P_{j,k} \nsim P_{i,\ell}$.

Observe that, under Definition 2.3, all adjacent branching points are unique. To illustrate this definition, consider the following example.

Example 2.1 (Adjacent branching point)

(i) $P_{2,2} \sim P_{1,1}$, but $P_{2,2} \nsim P_{1,2}$.
(ii) $P_{2,3} \sim P_{1,2}$, but $P_{2,3} \nsim P_{2,4}$.
(iii) $P_{3,5} \sim P_{2,3}$, but $P_{3,5} \nsim P_{1,1}$.
(iv) $P_{3,7} \sim P_{2,4}$, but $P_{3,7} \nsim P_{2,3}$.

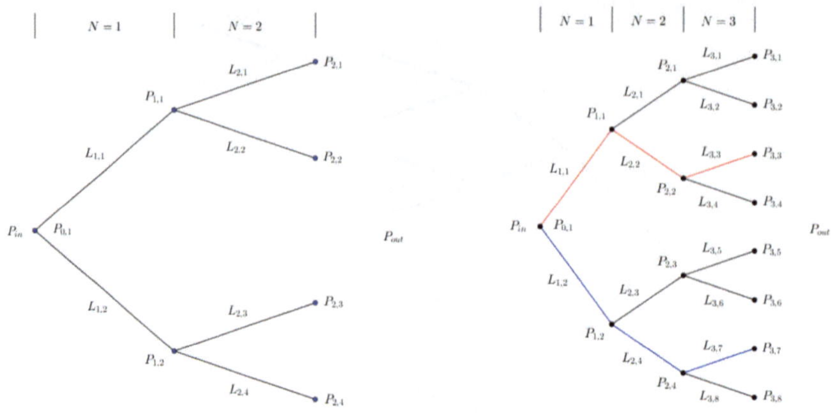

(a) $N = 2$. (b) $N = 3$.

Fig. 2.2 Self-similar network with potential-driven flow with **a** two and **b** three generations

Definition 2.4 (*Admissible paths*) Consider an N-generation network. An admissible path $\mathcal{P}_k$ is defined as the sequence of N adjacent branching points, starting from P_{in} and ending at P_{out}. In other words, the k-th admissible path in the network is described by

$$\mathcal{P}_k = \bigcup_{j=1}^{N} L_{N-j+1,\left\lceil \frac{k}{2^{j-1}} \right\rceil}, \quad 1 \leq k \leq 2^N. \tag{2.3}$$

The following example illustrates the previous definition.

Example 2.2 (Admisible path) For $N = 3$, there are 2^3 admissible paths. Let $k = 3$, then

$$\begin{aligned} \mathcal{P}_3 &= L_{3-1+1,\left\lceil \frac{3}{2^0} \right\rceil} \cup L_{3-2+1,\left\lceil \frac{3}{2^1} \right\rceil} \cup L_{3-3+1,\left\lceil \frac{3}{2^2} \right\rceil} \\ \mathcal{P}_3 &= L_{3,3} \cup L_{2,2} \cup L_{1,1}. \end{aligned} \tag{2.4}$$

In this vein, $\mathcal{P}_3$ is illustrated through the red path in Fig. 2.2b. Similarly, $\mathcal{P}_7$ is depicted through the blue path.

Note that in a two-generation network, for $N = 2$, there is 2^2 admisible paths, $2^3 - 2$ branches see Fig. 2.2a, the six transfer equations (one for each branch) and the conservation equations are:

$$\begin{aligned} L_{1,1}q_{1,1} &= P_{in} - P_{1,1}, & L_{2,2}q_{2,2} &= P_{1,1} - P_{out}, & & \\ L_{1,2}q_{1,2} &= P_{in} - P_{1,2}, & L_{2,3}q_{2,3} &= P_{1,2} - P_{out}, & q_{1,1} &= q_{2,1} + q_{2,2}, \\ L_{2,1}q_{2,1} &= P_{1,1} - P_{out}, & L_{2,4}q_{2,4} &= P_{1,2} - P_{out}. & q_{1,2} &= q_{2,3} + q_{2,4}. \end{aligned}$$

Taking into account the previous observations, we can deduce the following property.

Property 2.1 *Let $N \in \mathbb{N}$. Consider a network with $N-$generations and end points $P_{0,1} := P_{in}$ and $P_{N,k} := P_{out}$. Then, the transfer function of the $(j,k)-$branch is given by*

$$L_{j,k} q_{j,k} = \Delta P_{j,k}, \qquad 1 \leq k \leq 2^j, \quad 1 \leq j \leq N, \tag{2.5}$$

where $\Delta P_{j,k} := P_{j-1,\lceil \frac{k}{2} \rceil} - P_{j,k}$.

Proof Since the $(j,k)-$branch belongs to a self-similar network with potential-driven flow, the proof follows straightforwardly by noticing that the corresponding branch points are $P_{j-1,\lceil \frac{k}{2} \rceil}$ and $P_{j,k}$. □

Property 2.2 *For $j, k, N \in \mathbb{N}$, such that $1 \leq j \leq N$ and $1 \leq k \leq 2^N$, the sum of the transfer equations along any of the admissible paths is always $P_{in} - P_{out} = \Delta P$. That is, for any k, we have that*

$$\begin{aligned} \sum_{j=1}^{N} L_{N-j+1,\left\lceil \frac{k}{2^{j-1}} \right\rceil} q_{N-j+1,\left\lceil \frac{k}{2^{j-1}} \right\rceil} &= \sum_{j=1}^{N} \left(P_{N-j,\left\lceil \frac{k}{2^j} \right\rceil} - P_{N-j+1,\left\lceil \frac{k}{2^{j-1}} \right\rceil} \right) \\ &= P_{0,1} - P_{N,k} = P_{in} - P_{out} = \Delta P. \end{aligned} \tag{2.6}$$

Proof For any admissible path, the sum of the transfer equations is made through adjacent branching points. Expanding some terms of the sum, it is easy to see that it is a telescoping sum

$$(P_{N-1,\lceil \frac{k}{2} \rceil} - P_{N,\lceil \frac{k}{1} \rceil}) + (P_{N-2,\lceil \frac{k}{4} \rceil} - P_{N-1,\lceil \frac{k}{2} \rceil}) + \cdots + (P_{0,\lceil \frac{k}{2^N} \rceil} - P_{1,\lceil \frac{k}{2^{N-1}} \rceil}). \tag{2.7}$$

Here we use that $\lceil \frac{k}{2^N} \rceil = 1$, since $1 \leq k \leq 2^N$, so $\frac{k}{2^N} \leq 1$. Therefore, (2.7) reduces to $P_{0,1} - P_{N,k}$. □

Taking into account the Properties 2.1, 2.2, the transfer equations for a SSNPF with two-generation can be written down as:

$$\begin{aligned} L_{1,1}q_{1,1} + L_{2,1}q_{2,1} &= P_{in} - P_{out} = \Delta P, \qquad & L_{1,2}q_{1,2} + L_{2,3}q_{2,3} &= P_{in} - P_{out} = \Delta P, \\ L_{1,1}q_{1,1} + L_{2,2}q_{2,2} &= P_{in} - P_{out} = \Delta P, \qquad & L_{1,2}q_{1,2} + L_{2,4}q_{2,4} &= P_{in} - P_{out} = \Delta P. \end{aligned}$$

Now, by solving $q_{2,k}$ for $k \in \{1, \ldots, 4\}$, yields to

$$\begin{aligned} q_{2,1} &= L_{2,1}^{-1}(\Delta P - L_{1,1}q_{1,1}) + f(\phi_{L_{2,1}}), \qquad & q_{2,3} &= L_{2,3}^{-1}(\Delta P - L_{1,2}q_{1,2}) + f(\phi_{L_{2,3}}), \\ q_{2,2} &= L_{2,2}^{-1}(\Delta P - L_{1,1}q_{1,1}) + f(\phi_{L_{2,2}}), \qquad & q_{2,4} &= L_{2,4}^{-1}(\Delta P - L_{1,2}q_{1,2}) + f(\phi_{L_{2,4}}). \end{aligned}$$

Summing the previous equations so that we can apply the conservation equations

$$q_{1,1} = \left\{L_{2,1}^{-1} + L_{2,2}^{-1}\right\}\left(\Delta P - L_{1,1}q_{1,1}\right) + f(\phi_{L_{2,1}}, \phi_{L_{2,2}})$$
$$q_{1,2} = \left\{L_{2,3}^{-1} + L_{2,4}^{-1}\right\}\left(\Delta P - L_{1,2}q_{1,2}\right) + f(\phi_{L_{2,3}}, \phi_{L_{2,4}})$$

Now, we repeat the process to solve for the $q_{1,j}$'s

$$\left\{L_{2,1}^{-1} + L_{2,2}^{-1}\right\}^{-1} q_{1,1} = \Delta P - L_{1,1}q_{1,1} + \left\{L_{2,1}^{-1} + L_{2,2}^{-1}\right\}^{-1} f(\phi_{L_{2,1}}, \phi_{L_{2,2}}) - f(\phi_{L_{2,1}^{-1}+L_{2,1}^{-1}})$$
$$\left\{L_{2,3}^{-1} + L_{2,4}^{-1}\right\}^{-1} q_{1,2} = \Delta P - L_{1,2}q_{1,2} + \left\{L_{2,3}^{-1} + L_{2,4}^{-1}\right\}^{-1} f(\phi_{L_{2,3}}, \phi_{L_{2,4}}) - f(\phi_{L_{2,3}^{-1}+L_{2,4}^{-1}})$$

By combining like terms for the $q_{1,j}$'s, we obtain:

$$\begin{aligned} q_{1,1} = & \left(\left\{L_{2,1}^{-1} + L_{2,2}^{-1}\right\}^{-1} + L_{1,1}\right)^{-1} \Delta P + \left(\left\{L_{2,1}^{-1} + L_{2,2}^{-1}\right\}^{-1} + L_{1,1}\right)^{-1} \\ & \left(\left\{L_{2,1}^{-1} + L_{2,2}^{-1}\right\}^{-1} f(\phi_{L_{2,1}}, \phi_{L_{2,2}}) - f(\phi_{L_{2,1}^{-1}+L_{2,2}^{-1}})\right) + f(\phi_{(L_{2,1}+L_{2,2})+L_{1,1}}), \\ q_{1,2} = & \left(\left\{L_{2,3}^{-1} + L_{2,4}^{-1}\right\}^{-1} + L_{1,2}\right)^{-1} \Delta P + \left(\left\{L_{2,3}^{-1} + L_{2,4}^{-1}\right\}^{-1} + L_{1,2}\right)^{-1} \\ & \left(\left\{L_{2,3}^{-1} + L_{2,4}^{-1}\right\}^{-1} f(\phi_{L_{2,3}}, \phi_{L_{2,4}}) - f(\phi_{L_{2,3}^{-1}+L_{2,4}^{-1}})\right) + f(\phi_{(L_{2,3}+L_{2,4})+L_{1,2}}). \end{aligned}$$

By summing the two previous equations, we obtain:

$$q = \left[\left(\{L_{2,1}^{-1} + L_{2,2}^{-1}\}^{-1} + L_{1,1}\right)^{-1} + \left(\{L_{2,3}^{-1} + L_{2,4}^{-1}\}^{-1} + L_{1,2}\right)^{-1}\right] \Delta P + \sum \alpha_n \phi_n.$$

The assumptions on L consider that it has a unique inverse; therefore, the sum $\alpha_n \phi_n$ is indeed zero. Thus, we obtain a single equation.

$$\left[\left(\{L_{2,1}^{-1} + L_{2,2}^{-1}\}^{-1} + L_{1,1}\right)^{-1} + \left(\{L_{2,3}^{-1} + L_{2,4}^{-1}\}^{-1} + L_{1,2}\right)^{-1}\right]^{-1} q = \Delta P.$$

Therefore, the operator L_2, which describes the behavior of the two-generation network, is:

$$L_2 = \left[\left(\{L_{2,1}^{-1} + L_{2,2}^{-1}\}^{-1} + L_{1,1}\right)^{-1} + \left(\{L_{2,3}^{-1} + L_{2,4}^{-1}\}^{-1} + L_{1,2}\right)^{-1}\right]^{-1}.$$

For N generations, L_N can be expressed as:

$$L_N = \left\{ \sum_{j=1}^{2} \left[L_{1,j} + \left\{ \sum_{k=1}^{2} \left[L_{2,2(j-1)+k} + \left\{ \sum_{l=1}^{2} \left[L_{3,2(k-1)+l} + \cdots \right]^{-1} \right\}^{-1} \right]^{-1} \right\}^{-1} \right]^{-1} \right\}^{-1}$$

Our particular interest lies in self-similar networks where each bifurcation is governed by different operators. Specifically, replacing $L_{i,j}$ with L_1 and L_2 for odd and even j, respectively.

$$L_N = \left\{ \sum_{j=1}^{2} \left[L_j + \left\{ \sum_{k=1}^{2} \left[L_k + \left\{ \sum_{l=1}^{2} [L_l + \cdots]^{-1} \right\}^{-1} \right]^{-1} \right\}^{-1} \right]^{-1} \right\}^{-1}. \quad (2.8)$$

Proposition 2.1 (Equivalent operator for a network of N generations) *Consider a self-similar network with flow driven by a potential where L_1 and L_2 are linear operators with a unique inverse. Substituting $L_{i,j}$ with L_1 and L_2 for odd and even j, respectively, then the operator L_N inducing transfer along an N-generation network (from p_{in} to p_{out}) is*

$$L_N = \left\{ \sum_{j=1}^{2} \left[L_j + \left\{ \sum_{k=1}^{2} \left[L_k + \left\{ \sum_{l=1}^{2} [L_l + \cdots]^{-1} \right\}^{-1} \right]^{-1} \right\}^{-1} \right]^{-1} \right\}^{-1}. \quad (2.9)$$

Additionally, if we take their Laplace transforms $\mathscr{L}_{i,j} = \mathscr{L}[L_{i,j}]$ in the limit as $N \to \infty$, then

$$\mathscr{L}_\infty = \sqrt{\mathscr{L}_1 \mathscr{L}_2}. \quad (2.10)$$

Proof The original demonstration is detailed [5]. In our study, we conduct a specific analysis that delves deeply into the details concerning self-similar networks. We explore this concept thoroughly as it naturally leads to the application of fractional derivatives, which is pertinent to our application in the cardiovascular system. To clarify these ideas, we have included figures in the study that aid in understanding the core concepts of this mathematical model. If we consider the equations (2.8) and take their Laplace transforms, using the notation $\mathscr{L}_{i,j} = \mathscr{L}[L_{i,j}]$, then the equation can be written as:

$$\mathscr{L}_N = \cfrac{1}{\cfrac{1}{\mathscr{L}_1 + \cfrac{1}{\cfrac{1}{\mathscr{L}_1 + \cdots} + \cfrac{1}{\mathscr{L}_2 + \cdots}}} + \cfrac{1}{\mathscr{L}_2 + \cfrac{1}{\cfrac{1}{\mathscr{L}_1 + \cdots} + \cfrac{1}{\mathscr{L}_2 + \cdots}}}} \quad (2.11)$$

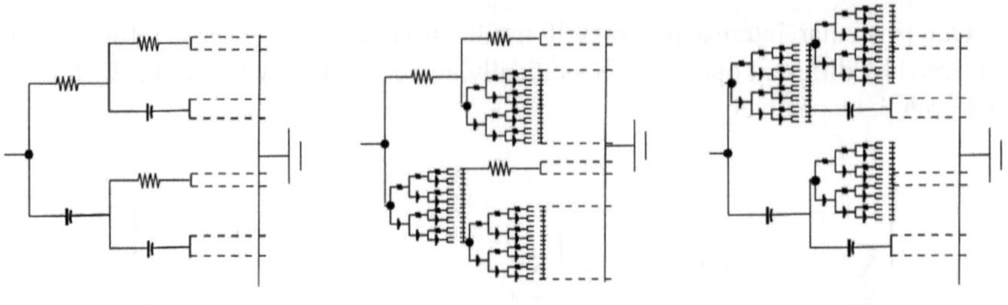

Fig. 2.3 Self-similar network electric circuit

In the limit as $N \to \infty$, it follows that:

$$\mathscr{L}_\infty = \frac{1}{\frac{1}{\mathscr{L}_1 + \mathscr{L}_\infty} + \frac{1}{\mathscr{L}_2 + \mathscr{L}_\infty}}, \quad \text{equivalent to} \quad \mathscr{L}_\infty = \sqrt{\mathscr{L}_1 \mathscr{L}_2}. \tag{2.12}$$

□

2.2.1 Self-similar Network Electric Circuit

The Proposition opens up the possibility of modeling a self-similar electrical circuit. A typical electrical circuit is composed of capacitors, resistors, and inductors, which have impedances $Z \propto (j\omega)^m$, where m is $-1, 0$, and $+1$, respectively. A device with fractance is one that does not fall into any of these three common categories, but rather has an impedance $Z \propto (j\omega)^\alpha$, with α being non-integer [14]. Refer to the Fig. 2.3.

For a branch with a resistor, the current-voltage relationship is given by:

$$Ri(t) = v(t), \tag{2.13}$$

and the branches with a capacitor are governed by:

$$\frac{1}{C} \int i(t)\, dt = \Delta v(t). \tag{2.14}$$

In the notation we have used before, $q = i$, $\Delta P = \Delta v$, for the resistances $L_1 = R$, for the capacitors $L_2 = \frac{1}{C} \int dt$. In the Laplace domain, $\mathscr{L}_1 = R$ and $\mathscr{L}_2 = 1/Cs$, such that

$$\mathscr{L}_\infty = \left(\frac{R}{C}\right)^{1/2} s^{-1/2}. \tag{2.15}$$

From Eq. (2.15), an interesting result can be deduced that relates voltage and current in a type of fractional Ohm's law. To achieve this, it is useful to keep in mind the definitions of the Riemann-Liouville fractional integral (RLI), the Caputo derivative, and the Laplace transform of the RLI. The proof of this theorem can be found in the book by Miller and Ross [15], in Chapter III, specifically in Eq. (6.5b).

Definition 2.5 (*Fractional Riemann-Liouville Integral of order alpha*) Let $\alpha > 0$ be any positive real number not integer, and let f be a locally absolutely integrable function. Let $\Gamma(\cdot)$ be the Euler gamma function. The Riemann-Liouville fractional integral of order α is defined as:

$$J^{\alpha} f(t) = \frac{1}{\Gamma(\alpha)} \int_0^t (t-\tau)^{\alpha-1} f(\tau)\, d\tau, \quad t > 0, \quad \alpha \in \mathbb{R}^+ \wedge \alpha \notin \mathbb{Z}. \tag{2.16}$$

Definition 2.6 (*Fractional Caputo Derivative of order alpha*) For $m \in \mathbb{Z}^+$, let $f \in \mathcal{AC}^m[0, T]$, $T > 0$, with an absolutely continuous derivative up to order $m-1$ and an absolutely continuous m-derivative. For $\alpha \in (m-1, m]$, the Caputo fractional derivative of order α, denoted by ${}^{C}D^{\alpha}$, is defined by ${}^{C}D^{\alpha}\ f(t) = J^{m-\alpha} D^m f(t)$, equivalent to

$${}^{C}D^{\alpha} f(t) = \frac{1}{\Gamma(m-\alpha)} \int_0^t (t-\tau)^{m-\alpha-1} f^{(m)}(\tau) d\tau, \tag{2.17}$$

where $f^{(m)}$ is the m-th order derivative of f.

Theorem 2.1 (Laplace transform of the Fractional Integral of Riemann-Liouville) *Let $\alpha > 0$ be any real number, $f \in AC[0, t]$. Furthermore, let f be of exponential order.*

$$\mathcal{L}[J^{\alpha}\ f(t)] = \frac{F(s)}{s^{\alpha}}, \quad (\mathcal{L}[f(t)] = F(s)\,). \tag{2.18}$$

The impedance as a function of ω is found by replacing s with $j\omega$ in the Eq. (2.15), then converting back from the Laplace domain to the natural domain; subsequently, contemplate the case where $\alpha = 1/2$. Note that the initial conditions $f^{(k)}(0)$ represent the derivatives of the function for integer orders. The operators employed in the analysis of self-similar trees are such that the initial conditions are zero. Finally, we present the fundamental theorem of calculus in the Caputo derivative version [16].

Theorem 2.2 (TFC Version Caputo) *Let $f(x)$ be a function $f : [a, b] \to \mathbb{R}$, with $-\infty < a < b < \infty$ and let $\alpha \in \mathbb{C}$ with $Re(\alpha) > 0$ and $n = \lceil Re(\alpha) \rceil$. If $f(x) \in AC^n[a, b]$ or $C^n[a, b]$, then, for $x \in (a, b)$:*

1. For $Re(\alpha) \notin \mathbb{N}$ or $\alpha \in \mathbb{N}$, we have

$$\left({}^{C}D^{\alpha} J^{\alpha}\ f\right)(t) = f(t).$$

2. *We have*

$$\left(J^{\alpha}\,{}^{C}D^{\alpha}\, f\right)(t) = f(t) - \sum_{j=0}^{n-1} \frac{(x-a)^j}{j!}\left(f^{(j)}(t)\Big|_{t=a}\right).$$

In view of the fundamental theorem of calculus in the Caputo version (Theorem 2.2), if the initial conditions of f are zero, the integral of Riemann-Liouville is the inverse of the Caputo derivative. In this sense, both definitions can be applied simultaneously. This means that the voltage $v(t)$ is the Riemann-Liouville fractional integral of the current $i(t)$ of a half order [17].

$$v(t) = \frac{\left(\frac{R}{C}\right)^{1/2}}{\Gamma(1/2)} \int_0^t \frac{i(\tau)}{(t-\tau)^{1/2}}\, d\tau. \tag{2.19}$$

The Eq. (2.19) can be expressed as $v(t) = (R/C)^{1/2} J^{1/2} i(t)$ in terms of the Eq. (2.16). Applying the fundamental theorem of calculus to the definition of the Riemann-Liouville operators, see [18]. The preceding equation also indicates that the current $i(t)$ is the Caputo derivative of order one-half of the voltage $v(t)$.

$$i(t) = \left(\frac{C}{R}\right)^{1/2} {}^{C}D^{1/2} v(t). \tag{2.20}$$

2.2.2 *Fractal Self-similar Networks Electric Circuit*

In this section, we will expand the modeling capabilities of self-similar networks. Let us consider a network constructed by replacing elements capacitors with self-similar networks, as depicted in the Fig. 2.3b. Analogous models have been employed to represent biological tissue [11]. We shall denote a network where capacitors have been replaced by self-similar networks as network (0, 1). The equivalent operator $\mathcal{L}_\infty = (R^3/C)^{1/4}/s^{1/4}$, resulting in a fractional derivative of order 1/4. In the case of (0, 2), that is, when capacitors have been replaced by self-similar networks, which in turn have capacitors replaced by networks, the equivalent operator is denoted as $\mathcal{L}_\infty = (R^7/C)^{1/8}/s^{1/8}$, resulting in a fractional derivative of order 1/8. In general, if a network is taken where $N-1$ capacitors are successively replaced by networks, such that the network has a depth of size N, then the equivalent operator will be $\mathcal{L}_\infty = \left(R^{2^{N+1}-1}/C\right)^{1/2^{N+1}} / s^{1/2^{N+1}}$, leading to the following voltage-current relation of order $1/2^{N+1}$

$$v(t) = \left(\frac{R^{2^{N+1}-1}}{C}\right)^{1/2^{N+1}} J^{1/2^{N+1}} i(t). \tag{2.21}$$

Consequently, it is possible to have fractional derivative orders of 1/2, 1/4, 1/8, and so forth.

Additionally, note that as the number $N \to \infty$, the equivalent operator $\mathscr{L}_\infty \to R$. This results in a purely resistive response.

On the other hand, consider constructing a network where resistive elements have been replaced by networks, see Fig. 2.3c. We will denote by (1,0) the network obtained in this manner; thus, the equivalent operator is $\mathscr{L}_\infty = (R/C^3)^{1/4}/s^{3/4}$. In the case where a network is constructed with resistive elements replaced by networks, which in turn have resistive elements replaced by networks; that is, the case (2,0), the equivalent operator is also $\mathscr{L}_\infty = (R/C^7)^{1/8}/s^{7/8}$.

For self-similar networks constructed with $M - 1$ resistive elements replaced by networks, resulting in a network with depth M, the equivalent operator is $\mathscr{L}_\infty = (R/C^{2^{M+1}-1})^{1/2^{M+1}}/s^{(2^{M+1}-1)/(2^{M+1})}$. Leading to a fractional derivative order of $1/2$, $3/4$, $7/8$, and so forth. If $M \to \infty$, $\mathscr{L}_\infty \to 1/(Cs)$, resulting in a purely capacitive response.

In general, for a network (M, N), $\mathscr{L}_\infty$ this leads to a fractional derivative order β defined by:

$$\beta = \frac{1}{2}\left(1 - \frac{1}{2^M} + \frac{1}{2^N}\right). \tag{2.22}$$

2.3 Cardiac Dynamics Modeled by Fractional Differential Equations

Modeling the cardiovascular system poses significant challenges, particularly in accurately representing both normal and pathological behaviors. For example, disruptions in what some studies [13, 19] describe as cardiac self-organization can lead to heart disease and failure. These disruptions can cause irregularities in electrical activity, as well as abnormal oscillations in left ventricular volume and pressure-manifesting as too rapid, too slow, or erratic rhythms.

Despite its limitations, the Windkessel model (WKM) remains a widely used lumped parameter approach within cardiac modeling, particularly effective for analyzing interactions following the insertion of a left ventricular assist device [12]. The WKM is represented by an electrical network, with its transfer function typically ranging from second to fourth order [20]. Instead of adding more elements to the WKM, an alternative approach involves incorporating a self-similar network of linear lumped elements, known as fractance elements [21], like those discussed in Sect. 2.2.2 and Eq. (2.22).

Briefly, we can consider the use of fractances in the left atrial, aortic and systemic compliances of the lumped parameter electric circuit. As mentioned in [21], the equations for left atrial pressure P_R, aortic pressure m and arterial pressure P_S from [13] are rewritten as follows:

$$\begin{aligned}
\frac{dV}{dt} &= \frac{1}{R_m}(P_R - P_V)\,\Theta(P_R - P_V) - \frac{1}{R_A}(P_V - m)\,\Theta(P_V - m) - \delta_p n,\\
\frac{d^{\alpha_m} m}{dt^{\alpha_m}} &= -\frac{1}{C_A}F_a + \frac{1}{C_A R_A}(P_V - m)\,\Theta(P_V - m) + \delta_p \frac{n}{C_A},\\
\frac{dF_a}{dt} &= \frac{m - P_S}{L_S} - \frac{R_C F_a}{L_S},\\
\frac{d^{\alpha_R} P_R}{dt^{\alpha_R}} &= \frac{-P_R + P_S}{C_R R_S} - \frac{1}{C_R R_M}(P_R - P_V)\,\Theta(P_R - P_V),\\
\frac{d^{\alpha_S} P_S}{dt^{\alpha_S}} &= \frac{P_R - P_S}{C_S R_S} + \frac{F_a}{C_S},
\end{aligned} \tag{2.23}$$

where $\alpha_m, \alpha_R, \alpha_S$ are the fractional orders of the time derivatives for m, P_R and P_S, respectively. V stands for the left ventricular volume, F_a is the aortic total flow, the function $\Theta(u)$ is a Heaviside function which takes the non-zero value of 1 for $u > 0$ or 0 otherwise. Moreover, the equation governing the pump flow is

$$\frac{dn}{dt} = \frac{1}{L_*}\left[P_V - m - R_* n + \beta\omega^2\right], \tag{2.24}$$

where ω is the rotational speed of the pump. L_*, R_* and β are constants taken from reference [12]. To effectively analyze cardiac muscle dynamics, it is essential to study pressure-volume loops. Interestingly, the system described by Eqs. (2.23) and (2.24) demonstrates both normal and pathological behaviors, including conditions such as mitral regurgitation, stenosis, and others. See Fig. 2.4 for examples of PV loops obtained by varying the fractional order of the model's fractances.

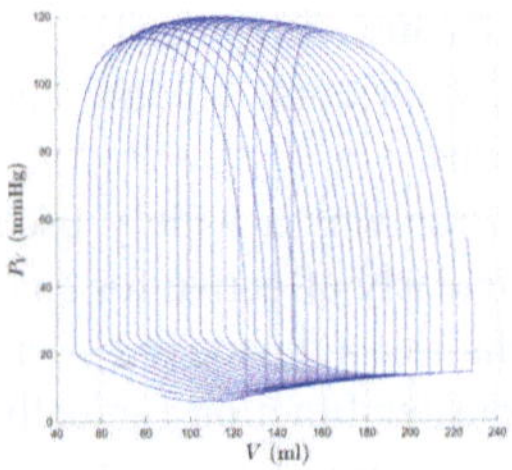

(a) $\alpha_s = 0.2$, $\alpha_m = \alpha_r = 1$.

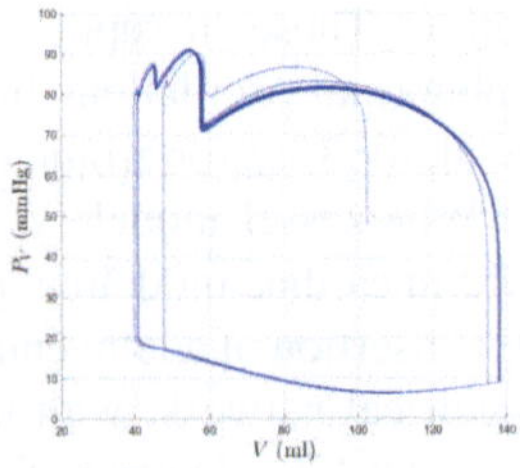

(b) $\alpha_m = 0.6786$, $\alpha_s = \alpha_r = 1$

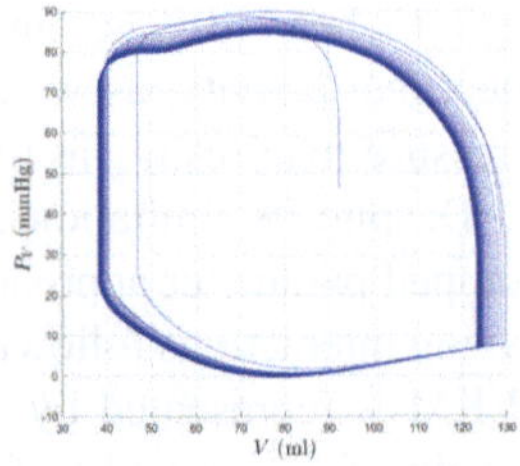

(c) $\alpha_r = 0.4631$, $\alpha_s = \alpha_m = 1$

Fig. 2.4 Pressure-volume loop graphs for different values of the derivative order in Eq. (2.23)

2.4 Conclusions

In this document, we present a derivation, distinct from classical approaches, to arrive at the fractional differential and integral operators, such as the Laurent derivation based on Cauchy's nth derivative theorem for a function of a complex variable, or through the approach that extends the use of the Gamma function in iterated integrals. The concept of self-similar networks has been explored by Mihir Sen and collaborators [4, 5], and here we offer a formal introduction from a mathematical perspective-or at least that was the intention. Similarly, our goal is to contribute to the literature with an alternative way of presenting fractional operators and, therefore, to encourage the mathematical modeling of complex phenomena using these types of networks. Moreover, we present a highly complex model that is naturally interesting due to its implications for human health. Thus, based on self-similar networks, we provide an analysis of electrical circuits with fractal elements, allowing us to model the dynamics of cardiac muscle through a hybrid system of fractional and ordinary differential equations. The main result of the model thus constructed is that it enables the recovery of pressure-volume loops that characterize both pathological and normal behaviors, offering an advantage over models that rely solely on integer-order derivatives.

Acknowledgements JEEM extends its sincere gratitude to Marisol Martínez-Bello for her invaluable contributions and insightful discussions on the cardiovascular system. AJGC wishes to express deep appreciation to Professors Mihir Sen and Eun-jin Kim for their guidance and introduction to the problems of self-similar networks and the cardiovascular system, respectively. The work of J.-E. Escalante-Martínez and C.-F. Méndez-Barrios has been supported by CONAHCyT–México, under the grant *Ciencia Básica y de Frontera:* CBF2023–2024–2696.

Competing Interests The authors have no conflicts of interest to declare that are relevant to the content of this chapter.

References

1. Boccaletti S, Pisarchik AN, Del Genio CI, Amann A (2018) Synchronization: from coupled systems to complex networks. Cambridge University Press
2. Cheng X, Scherpen JMA (2021) Model reduction methods for complex network systems. Ann Rev Control Robot Autonom Syst 4(1):425–453
3. McPhee JJ (1996) On the use of linear graph theory in multibody system dynamics. Nonlinear Dyn 9:73–90
4. Mayes J, Sen M (2011) Approximation of potential-driven flow dynamics in large-scale self-similar tree networks. Proc Royal Soc A Mathem Phys Eng Sci 467(2134):2810–2824
5. Sen M, Hollkamp JP, Semperlotti F, Goodwine B (2018) Implicit and fractional-derivative operators in infinite networks of integer-order components. Chaos, Solitons Fractals 114:186–192
6. Ni X, Goodwine B (2022) Frequency response and transfer functions of large self-similar networks. J Dyn Syste Measur Control 144(8):081007
7. Guel-Cortez A-J, Méndez-Barrios C-F, Kim E, Sen M (2021) Fractional-order controllers for irrational systems. IET Control Theory Appl 15(7):965–977

8. Goodwine B (2023) Fractional-order dynamics in large scale control systems. In: 2023 31st Mediterranean conference on control and automation (MED), pp 747–752. IEEE
9. Sabatier J (2022) Probabilistic interpretations of fractional operators and fractional behaviours: extensions, applications and tribute to prof. josé tenreiro machado's ideas. Mathematics 10(22):4184
10. Hristov J (2023) Constitutive fractional modeling. Contemporary mathematics. mathematical modelling: principle and theory 786:37–140
11. Kelly JF, MacGough RJ (2009) Fractal ladder model and power law wave equations. J Acoust Soc America 126(4):2072–2081
12. Simaan MA, Ferreira A, Chen S, Antaki JF, Galati DG (2009) A dynamical state space representation and performance analysis of a feedback-controlled rotary left ventricular assist device. IEEE Trans Control Syst Technol 17(1)
13. Kim EJ, Capoccia M (2020) Mechano-electric effect and a heart assist device in the synergistic model of cardiac function. Mathem Biosciences Eng 17(5):5212–5233
14. Nakagawa M, Sorimachi K (1992) Basic characteristics of a fractance device. IEICE Trans Fundamentals Electron Commun Comput Sci 75(12):1814–1819
15. Miller KS, Ross B (1993) An introduction to the fractional calculus and fractional differential equations. Wiley
16. Grigoletto EC, de Oliveira EC (2013) Fractional versions of the fundamental theorem of calculus. Appl Mathem
17. Herrmann R (2011) Fractional calculus: an introduction for physicists. World Scientific
18. Grigoletto EC, de Oliveira EC (2013) Fractional versions of the fundamental theorem of calculus. Appl Mathem
19. Kim E, Capoccia M (2019) Synergistic modelling of cardiac function with a heart assist device. Phys Rev E
20. Capoccia M (2015) Development and characterization of the arterial w indkessel and its role during left ventricular assist device assistance. Artif Organs 39(8):E138–E153
21. Guel-Cortez A-J, Kim E-J (2021) A fractional-order model of the cardiac function. In: Skiadas CH, Dimotikalis Y (eds) 13th Chaotic modeling and simulation international conference, pp 273–285, Cham, 2021. Springer International Publishing

Chapter 3
Fractional Derivatives and Hopf Bifurcation in Ecological Interactions Among Plants, Pollinators, and Herbivores

Jesús Enrique Escalante-Martínez, Porfirio Toledo, and José Alfredo Zavaleta-Viveros

3.1 Introduction

In ecosystems, interactions between species are essential to maintain ecological balance. These relationships greatly influence species' survival and evolution patterns, ranging from mutualism to predation [25]. The mutualistic interaction between pollinators and plants and the predator-prey dynamics between herbivores and plants deserve a detailed analysis because of their ecological [2, 14, 23].

The development of mathematical modeling of these interactions has made remarkable advances since the work of Holling, who introduced functional response functions to describe predator-prey relationships [8]. These functions have been fundamental to understanding the dynamics of the population and their interactions. However, models based on traditionally used differential equations of integer order have limitations, particularly their inability to incorporate memory effects and complex temporal dependencies inherent in biological systems.

In this context, fractional derivatives are a very useful tool as they allow us to consider the idea of memory and cumulative effects. Mathematical models represented as systems of fractional-order differential equations have been proven to be effective in the representation and description of biological processes that involve complex temporal interactions, such as the study of metabolic rhythms, disease propagation, and population dynamics [1, 6, 7, 17].

J. E. Escalante-Martínez
Facultad de Ingeniería Mecánica y Eléctrica, Universidad Veracruzana, Poza Rica, Veracruz, México
e-mail: jeescalante@uv.mx

P. Toledo (✉) · J. A. Zavaleta-Viveros
Facultad de Matemáticas, Universidad Veracruzana, Xalapa, Veracruz, México
e-mail: ptoledo@uv.mx

J. A. Zavaleta-Viveros
e-mail: josezavaleta@uv.mx

J. J. Rosales García et al. (eds.), *Advances in Fractional Calculus*,
Modeling and Optimization in Science and Technologies 21,
https://doi.org/10.1007/978-3-031-83989-4_3

In this paper, we propose an approach to analyzing pollinator-plant-herbivore interactions that employs fractional-order derivatives. This approach not only captures the nonlinear and time-dependent dynamics of these systems, but also provides a more complete understanding of the underlying mechanisms governing these interactions, thus contributing to knowledge in the fields of biodiversity and ecosystem sustainability.

3.2 Pollinator-Plant-Herbivore Model

3.2.1 *Integer-Order Pollinator-Plant-Herbivore Model*

Interactions of great importance for ecosystems occur between pollinators and plants, such as seed dispersal, crop pollination, and air purification. Mutual benefits between these populations arise each time the pollinator visits the plants. While the pollinator receives its benefit in the form of food by extracting nectar from the flowers, plants benefit from the simple fact that they are pollinated. Each time the pollinator visits the plant, it is impregnated with pollen, which is transported to other places, contributing to the plant's propagation. The role played by a third population, herbivores, interacting with two populations that maintain a mutualistic interaction is a type of predation. The existence of predators in any ecosystem is essential because their presence helps to control prey populations. If we consider the herbivore a plant predator, we can say that control occurs naturally.

The plant-herbivore interactions that we consider as herbivore satiation [15], at the same time, may reflect a defense strategy on the part of plants [26]. The rate of change of these populations is a function that varies in terms of events that occurred in some previous time interval; that is, they are functions with memory.

In [24] a mutualism model is proposed between a pollinator population and a plant population. The transfer of pollen by pollinators is beneficial to plants, while plants provide an energetic reward to the pollinator by providing nectar and pollen. Consider that $y(t)$ represents the plant population, while $x(t)$ represents the pollinator population.

The equation for the growth of $y(t)$ is as follows:

$$\frac{dy}{dt} = \frac{k_1 \sigma \mu x y}{1 + \sigma \phi \mu^2 y} - \gamma y. \tag{3.1}$$

The first term represents the plant birth rate and is based on a type II Holling functional response, interpreting pollinators as plant predators. The coefficient k_1 represents the efficiency constant. The attack rate is interpreted as σ, where σ is the probability of encounters and μ is the energetic reward. The manipulation time is modeled as ϕ, where ϕ is the speed of nectar extraction. In the second term γ is the mortality rate of the plant population.

The population growth of $x(t)$ is modeled as

$$\frac{dx}{dt} = x(\rho - bx) - \left(\lambda x - \frac{k_2 \sigma \mu^2 xy}{1 + \sigma \phi \mu^2 y}\right). \tag{3.2}$$

The first term is the pollinator birth rate, where ρ is the maximum birth rate and b is the regulation counter. The second term is the pollinator death rate, where λ is the maximum death rate in the absence of the plant population and k_2 is the energy conversion constant. The constant $k = \frac{\rho - \lambda}{b}$ will represent the carrying capacity and is interpreted as a measure of the degree of specialization of the pollinator on the plant, the Eq. (3.2) transforms:

$$\frac{dx}{dt} = bx(k - x) + \frac{k_2 \sigma \mu^2 xy}{1 + \sigma \phi \mu^2 y}. \tag{3.3}$$

For the system of Eqs. (3.1) and (3.3), we have that when $k = 0$ the pollinator population does not survive without that of the plants; while when k becomes very large, the absence of plants has no effect on the pollinator population, see [9, 24]. It is also shown that, under certain conditions, the solutions to the system are nonnegative and bounded.

In [9, 20] a population of herbivores is introduced into the system formed by (3.1) and (3.3). The herbivore population is assumed to consume the plant flowers, so that the plants provide a lower energy reward to the pollinators, which is detrimental to the plant population. If $z(t)$ represents the herbivore population and the herbivores are considered to be predators of the plants; using a type II Holling functional response, the system of equations posed for the three populations is as follows:

$$\begin{aligned}
\frac{dx}{dt} &= bx(k - x) + g(z)\frac{k_2 \sigma \mu^2 xy}{1 + \sigma \phi \mu^2 y},\\
\frac{dy}{dt} &= g(z)\frac{k_1 \sigma \mu xy}{1 + \sigma \phi \mu^2 y} - \gamma y - \frac{m_1 yz}{a + y},\\
\frac{dz}{dt} &= \frac{m_2 yz}{a + y} - \delta z,
\end{aligned} \tag{3.4}$$

where a is the saturation constant, m_1 is the maximum ingestion rate, m_2 is the maximum growth rate of the herbivore (with $0 < m_2 \leq m_1$), and δ is the death rate of $z(t)$; and $g(z)$ is the amount by which the pollinator visitation rate is reduced, $g \in C^1[0, +\infty)$, $g(0) = 1$, $g'(z) \leq 0$ and $g(z) > 0$, for all $z \geq 0$.

Note that if $g(z) = 1$ for all $z \geq 0$, then the conditions imposed for $g(z)$ are satisfied, and it follows that the herbivore response does not affect the pollinator visitation rate and therefore there is no effect on the plant growth rate. In [9] it is shown that the solutions of system (3.4) are nonnegative and bounded.

In [21] the Holling functional response is changed from type II to type Holling IV in the way $y(t)$ and $z(t)$ interact, proposing the model

$$
\begin{aligned}
\frac{dx}{dt} &= bx(k-x) + g(z)\frac{k_2\sigma\mu^2 xy}{1+\sigma\phi\mu^2 y}, \\
\frac{dy}{dt} &= g(z)\frac{k_1\sigma\mu xy}{1+\sigma\phi\mu^2 y} - \gamma y - \frac{m_1 yz}{\frac{y^2}{c}+y+a}, \\
\frac{dz}{dt} &= \frac{m_2 yz}{\frac{y^2}{c}+y+a} - \delta z.
\end{aligned}
\tag{3.5}
$$

When $c \to \infty$ we have $\frac{y^2}{c} \to 0$ and the Holling functional response type IV is seen as type II.

3.2.2 *Basic Concepts of Fractional Calculus*

While rooted in historical context, the terms fractional derivatives and fractional integrals in calculus can stated more accurately as differential and integral operators of arbitrary order [11, 18, 22]. Various authors developed definitions of the fractional derivative to improve mathematical modeling and analysis of complex systems. Among these, the Riemann-Liouville definition, which emerged from the development of classical integral theories, stands out for its wide application in various fields of science and engineering [10].

Definition 3.1 The Riemann-Liouville fractional integral of order α, with $\alpha \in \mathbb{R}^+$, for a function $f : \mathbb{R}^+ \to \mathbb{R}$ is defined as:

$$
J^\alpha f(t) = \frac{1}{\Gamma(\alpha)} \int_0^t (t-\tau)^{\alpha-1} f(\tau)\, d\tau. \tag{3.6}
$$

$\Gamma(\cdot)$ denotes the Gamma function.

Definition 3.2 Let f be an absolutely locally integrable function on $\mathbb{R}^+$, moreover causal; that is, identically zero for $t < 0$. For $\alpha \in (m-1, m]$, with $m \in \mathbb{Z}^+$, the Riemann-Liouville fractional derivative of order α, denoted by D^α, is defined by

$$
D^\alpha f(t) := D^m J^{m-\alpha} f(t), \tag{3.7}
$$

equivalent to,

$$
D^\alpha f(t) = \frac{d^m}{dt^m}\left(\frac{1}{\Gamma(m-\alpha)} \int_0^t (t-\tau)^{m-\alpha-1} f(\tau)\, d\tau\right). \tag{3.8}
$$

A limitation of this definition is that the derivative of a constant is not always zero. To address this, Michele Caputo proposed a definition ensuring that the fractional derivative of any constant equals zero [3, 10].

Definition 3.3 For $m \in \mathbb{Z}^+$, let $f \in \mathcal{AC}^m[0, T]$, $T > 0$, with an absolutely continuous derivative up to order $m - 1$ and an absolutely continuous m-derivative. For $\alpha \in (m - 1, m]$, the Caputo fractional derivative of order α, denoted by ${}^C D^\alpha$, is defined by

$$ {}^C D^\alpha \; f(t) = J^{m-\alpha} D^m f(t), \tag{3.9}$$

equivalent to,

$$ {}^C D^\alpha f(t) = \frac{1}{\Gamma(m-\alpha)} \int_0^t (t-\tau)^{m-\alpha-1} f^{(m)}(\tau) d\tau, \tag{3.10}$$

where $f^{(m)}$ is the m-th order derivative of f.

The primary distinction between Caputo's and Riemann-Liouville's definitions lies in their behavior under the Laplace transform. Caputo's definition involves only integer-order derivatives in the transform, whereas Riemann-Liouville's reintroduces fractional derivatives [13]. However, the functional domain for Caputo's derivative is more restrictive compared to that for Riemann-Liouville [12]. An intriguing example of a real-valued function that is uniformly continuous but not absolutely continuous can be found in [5]. For insights into the physical and geometrical interpretations of fractional derivatives, Podlubny's work is instructive [19].

The formulation of Caputo's derivative, as seen in Eq. (3.10), implies that the dynamics at time t depend on the entire history of events from the initial time t_0 to t, thereby encapsulating the memory of the modeled phenomenon; this makes it a particularly suitable choice for modeling complex ecological interactions.

The following stability result is fundamental in this context [4, 16].

Theorem 3.1 *Consider the fractional-order system given below:*

$$ {}^C D^\alpha x(t) = f(x(t)), \; x(0) = x_0 \in \mathbb{R}^n, \; 0 < \alpha < 1, \tag{3.11}$$

where $x(t) = (x_1(t), x_2(t), \ldots, x_n(t)) \in \mathbb{R}^n$ *and* $f : (f_1, f_2, \ldots, f_n) : \mathbb{R}^n \to \mathbb{R}^n$. *The equilibrium points of the system* (3.11) *are determined by solving equation* $f(x) = 0$.

We suppose x^* *is an equilibrium point of system* (3.11) *and the Jacobian matrix* $J = \frac{\partial f}{\partial x}$. *The equilibrium point* x^* *is locally asymptotically stable, for all eigenvalues* λ_i *of* J, *if and only if* $\min\{|\arg(\lambda_i)| : 1 \le i \le n\} > \frac{\alpha\pi}{2}$. *The equilibrium* x^* *is unstable when the following condition holds* $\min\{|\arg(\lambda_i)| : 1 \le i \le n\} \le \frac{\alpha\pi}{2}$.

3.2.3 Fractional-Order Pollinator-Plant-Herbivore Model

Let $x(t)$, $y(t)$, and $z(t)$ represent the populations of pollinators, plants, and herbivores, respectively. If we consider the system defined by (3.5) with the assumption that $g(z) \equiv 1$, then the conditions imposed on $g(z)$ are satisfied, and it is additionally interpreted that the herbivore response does not affect the pollinator visitation rate, and consequently there is no effect on the plant growth rate.

For this analysis, we will consider the following system of fractional differential equations:

$$
\begin{aligned}
{}^{C}D^{\alpha}x &= bx(k-x) + \frac{k_2\sigma\mu^2 xy}{1+\sigma\phi\mu^2 y}, \\
{}^{C}D^{\alpha}y &= \frac{k_1\sigma\mu xy}{1+\sigma\phi\mu^2 y} - \gamma y - \frac{m_1 yz}{\frac{y^2}{c}+y+a}, \\
{}^{C}D^{\alpha}z &= \frac{m_2 yz}{\frac{y^2}{c}+y+a} - \delta z,
\end{aligned}
\tag{3.12}
$$

where $x(0) \geq 0$, $y(0) \geq 0$, and $z(0) \geq 0$. The parameters in (3.12) are positive real numbers, each carrying specific ecological significance.

3.3 Parameter Analysis

3.3.1 Comparison of the Results with the Case of Integer Derivatives

Based on the numerical phase space analysis of the system (3.12) for the case $\alpha = 1$, as discussed in [21], the following parameters were considered: $a = 1, b = 1, c = 4$, $k = 2$, $k_1 = 1$, $k_2 = 3.5$, $m_1 = 2$, $m_2 = 3$, $\gamma = 2.5$, $\delta = 0.45$, $\phi = 1$, $\mu = 1$ and $\sigma = 2$.

We can now identify the equilibrium points of this system, which are listed in Table 3.1. It is important to note that only one of them has positive coordinates.

Given the only plausible equilibrium point $P = (2.9183, 0.1778, 1.0702)$, we observe that the Jacobian matrix $J(P)$ is

$$
J(P) = \begin{bmatrix} -2.9183 & 11.1145 & 0 \\ 0.2623 & -0.8347 & -0.3 \\ 0 & 2.2655 & 0 \end{bmatrix}. \tag{3.13}
$$

The characteristic polynomial of the Jacobian matrix is $p(\lambda) = \lambda^3 + 3.7531\lambda^2 + 0.1995\lambda + 1.9835$. Solving the equation $p(\lambda) = 0$ we obtain the eigenvalues $\lambda_1 = -3.8359$, $\lambda_2 = 0.0413 + 0.7178\,i$ and $\lambda_2 = 0.0413 - 0.7178\,i$. These values represent the system's stability and give important information about its dynamic behavior.

Note that $\min\{|\arg(\lambda_i)| : 1 \leq i \leq n\} = 0.4816\pi$ and, according to Theorem 3.1, the point P will be locally asymptotically stable if $\alpha \in (0, 0.9632)$; while P will be unstable if $\alpha \in [0.9632, 1)$.

Table 3.1 Equilibrium points of the system (3.12)

x	y	z	Plausible
6.6666	−2.0	0	×
0	22.4888	−187.4066	×
0	0.1778	−1.4822	×
0.9664	−0.1141	0	×
2.9183	0.1778	1.0702	✓
0	0	0	×
5.4238	22.4888	−169.7203	×
0	−0.5	0	×
2	0	0	×
0	−2.0	0	×
4.5353	1.3141	0	×

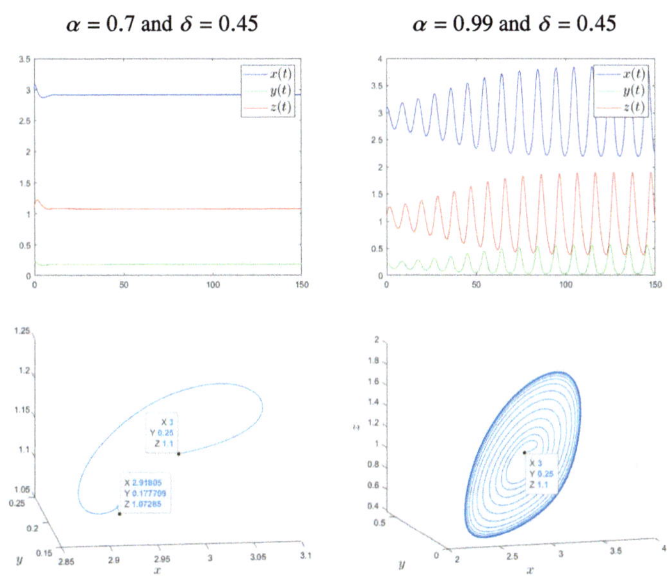

Fig. 3.1 The system trajectories for $\delta = 0.45$, with an initial condition of $(3, 0.25, 1.1)$, with $\alpha = 0.7$ and $\alpha = 0.99$

Figure 3.1 illustrates the behavior of the system for $\delta = 0.45$ with an initial condition of $(3, 0.25, 1.1)$ and values of α in the intervals $(0, 0.9632)$ and $[0.9632, 1)$. The system trajectories converge to the equilibrium point for $\alpha \in (0, 0.9632)$, indicating asymptotic stability. This behavior indicates that, in this case, the interactions between the populations reach an equilibrium state over time. In contrast, for $\alpha \in [0.9632, 1)$, the trajectories do not converge toward the equilibrium point; instead, they move away from it and tend toward a possible attractor cycle.

This phenomenon indicates a qualitative change in the system's behavior, in which the populations begin to persistently oscillate rather than stabilize. These periodic oscillations indicate the presence of a limit cycle resulting from changes in the order of differentiation. This qualitative change in system behavior has significant implications for population dynamics. While in the case of $\alpha \in (0, 0.9632)$, populations tend toward static equilibrium, in the case of $\alpha \in [0.9632, 1)$, populations experience continuous oscillations, which could correspond to fluctuations in population densities due to environmental factors or complex biological interactions.

3.3.2 *Numerical Simulations*

This section shows the results of various numerical simulations to study the system's dynamics around a bifurcation point. As in the previous section, for the present simulations, we use the parameters considered for the case of order $\alpha = 1$ in [21], except for the parameter δ, which is subject to variation. In addition to the above, we are interested in seeing the behavior of the system dynamics concerning changes in the derivative order α. The results show a change in the stability of the equilibrium point and the appearance of periodic oscillations, indicating the possible existence of a Hopf bifurcation point.

We can observe in Figs. 3.1 and 3.2, that for $\alpha = 0.99$, if δ takes different values from 0.3 to 0.6, the equilibrium point transitions from unstable to stable and the cycle disappears. Assuming that the eigenvalues of the characteristic polynomial change continuously with respect to the δ parameter, there must exist a value for which the real parts of the eigenvalues cancel out, indicating the presence of a Hopf bifurcation, where the system transitions from having oscillations to a stable equilibrium point. This behavior is analogous to that observed in the case of order $\alpha = 1$, discussed in [21].

On the other hand, when α takes smaller values, no limit cycles are present, but the equilibrium point remains as an attractor even for small values of δ, as can be seen in Figs. 3.1 and 3.2, this shows the sensitivity of the system (3.12) with respect to the derivative order α, suggesting that the dynamics of the system is highly order dependent. Note that the interval in which α takes values for which the limit cycle does not appear is greater than the interval in which it does appear. The latter indicates that when we reduce the order of the fractional derivative, the system tends to behave predictably and stably, decreasing the probability of cyclic oscillations.

3.4 Conclusions

The results observed in the simulations highlight the importance of the order of the fractional derivative in the dynamics of the modeled ecological systems. While for values close to $\alpha = 1$, the system may experience bifurcations that introduce limit

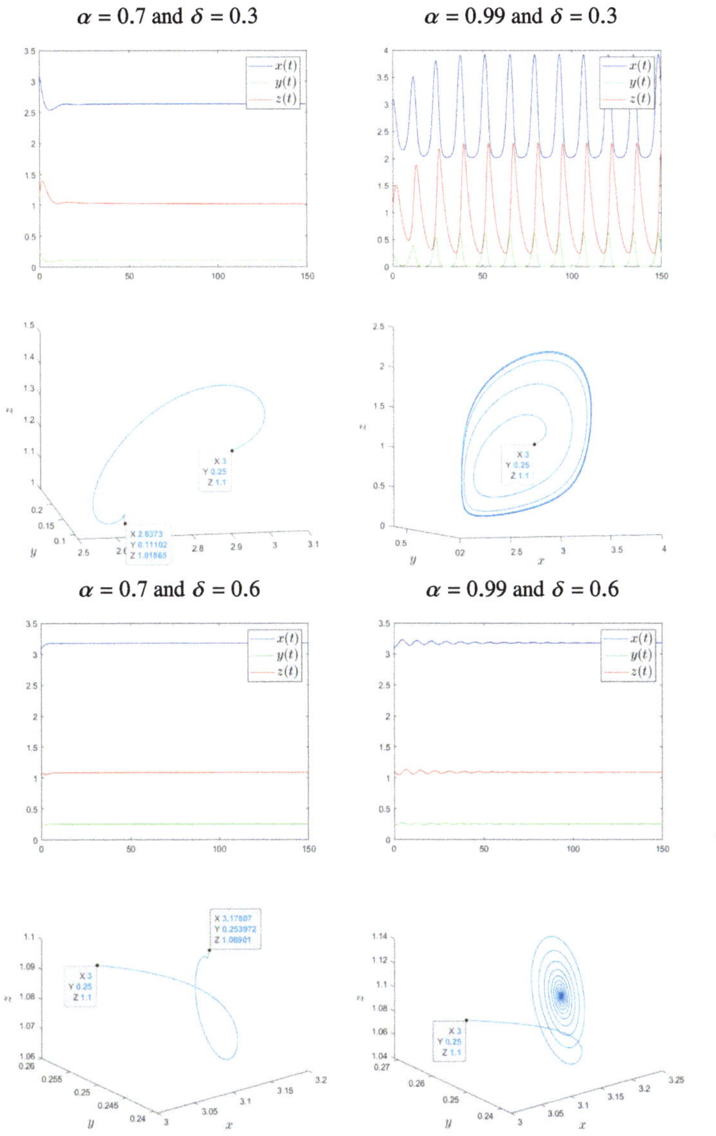

Fig. 3.2 The system trajectories for $\delta = 0.3$ and $\delta = 0.6$, with $\alpha = 0.7$ and $\alpha = 0.99$, starting from the initial condition $(3, 0.25, 1.1)$

cycles, for lower values of α, the system exhibits a more robust tendency toward stability at the equilibrium point, which could have significant implications for the modeling and control of complex systems in ecology and other disciplines.

According to the numerical simulations performed, and considering that δ represents the mortality rate of the herbivore population, it is observed that as this rate increases, the system becomes less susceptible to periodic fluctuations, reaching a state of stable equilibrium. The existence of a critical value of δ, where the real parts of the eigenvalues vanish, would mark the threshold at which the system transitions from an oscillatory state, characterized by undesired periodic fluctuations in the population, to a stable equilibrium state, where the oscillations extinguish.

On the other hand, considering that the system is sensitive to the derivative order and considering that we can interpret δ as a population control or management parameter, where we use an induced mortality rate to maintain the population within certain desirable limits, the above analysis shows that this parameter can be a crucial management tool to control population stability. In systems where no limit cycle occurs, management could focus on maintaining δ at levels that favor stability, while in systems with possible oscillations, adjusting δ could prevent unwanted fluctuations; this has practical applications in the management of nature reserves or in conservation programs where the goal is to maintain populations within sustainable limits, avoiding both overpopulation and extinction.

In the traditional framework, the system's state is determined instantaneously, without consideration of temporal windows or the past behavior of the ecosystem under investigation. In contrast, the fractional framework considers the system's historical evolution to be of fundamental importance in determining its state at any given time. Modifications to the parameter δ may indicate substituting one herbivore species with another, displaying differences in lifespan, or reflecting the acquisition of resistance to a chemical agent by a species, thereby increasing its longevity. Conversely, alterations in the differentiation order indicate the system's resilience or capacity to restore equilibrium following environmental perturbations. Systems with orders closer to integers exhibit reduced memory effects, while those with orders distant from integers retain more extensive memory. More stable ecosystems rely on more comprehensive historical data to determine their current state. Human interventions often disrupt these natural processes, leading to a scenario where the ecosystem is assessed immediately (classical case, integer derivative), neglecting its historical context (non-integer case, fractional derivative). This observation aligns with empirical evidence, where disturbed ecosystems are more prone to undesirable dynamic behaviors, such as significant fluctuations in species populations, which can lead to species extinction.

A future task is to provide a proof of the existence of the limit cycle and the Hopf bifurcation. This can be done by investigating the stability crossing limit and arguing for the continuity of the roots of the characteristic polynomial. In addition, it is necessary to extend the parameter analysis to show stability regions. Multiple factors can increase the mortality rate of the three species; that is, in a real system, many unmodeled factors can reduce the lifetime of any of the three species involved. It is important to determine the fragility of the system stability, as these rates vary.

Acknowledgements JEEM would like to thank Marisol Martínez-Bello for her support in making this research possible, as well as for contributing her expertise to many discussions on ecological interactions.

Competing Interests The authors have no conflicts of interest to declare that are relevant to the content of this chapter.

References

1. Alzahrani EO, Khan MA (2018) Modeling the dynamics of Hepatitis E with optimal control. Chaos Solitons Fract 116:287–301 Nov
2. Aronson JC, Blatt CM, Aronson (2016) Restoring ecosystem health to improve human health and well-being: physicians and restoration ecologists unite in a common cause. Ecol Soc 21(4):art39
3. Caputo M (1969) Elasticità e dissipazione. Zanichelli, Bologna
4. Cui Z, Zhou Y, Li R (2023) Complex dynamics analysis and chaos control of a fractional-order three-population food chain model. Fractal Fract 7:7
5. Dovgoshey O, Martio O, Ryazanov V, Vuorinen M (2006) The Cantor function. Expositiones Mathematicae 24(1):1–37 Feb
6. Escalante-Martínez JE, Gómez-Aguilar JF, Calderón-Ramón C, Aguilar-Meléndez A, Padilla-Longoria P (2018) A mathematical model of circadian rhythms synchronization using fractional differential equations system of coupled van der Pol oscillators. Int J Biomathem 11(01):1850014 Jan
7. Ghanbari B, Gómez-Aguilar JF (2018) Modeling the dynamics of nutrient-phytoplankton-zooplankton system with variable-order fractional derivatives. Chaos Solitons Fractals 116:114–120 Nov
8. Holling CS (1965) The functional response of predators to prey density and its role in mimicry and population regulation. Memoirs Entomol Soc Canada 97(S45):5–60 May
9. Jang SRJ (2002) Dynamics of herbivore-plant-pollinator models. J Mathem Biol 44(2):129–149 Feb
10. Kilbas AA, Srivastava HM, Trujillo JJ (2006) Theory and applications of fractional differential equations. Elsevier B.V, Amsterdam
11. Lazarević MP, Rapaić MR, Šekara TB (2014) Introduction to fractional calculus with brief historical background. In: Mladenov V, Mastorakis N (eds) Advanced topics on applications of fractional calculus on control problems, system stability and modeling, chapter 1, pp 3–15. WSEAS Press
12. Li C, Qian D, Chen Y (2011) On Riemann-Liouville and Caputo derivatives. Discrete Dyn Nature Soc 1–15:2011
13. Liang S, Wu R (2015) Laplace transform of fractional order differential equations. Electron J Differential Equ 2015(139):1–15
14. Liss KN, Mitchell MGE, MacDonald GK, Mahajan SL, Méthot J, Jacob AL, Maguire DY, Metson GS, Ziter C, Dancose K, Martins K, Terrado M, Bennett EM (2013) Variability in ecosystem service measurement: a pollination service case study. Frontiers Ecol Environ 11(8):414–422 Oct
15. Lucas-Barbosa D, Van Loon JJA, Dicke M (2011) The effects of herbivore-induced plant volatiles on interactions between plants and flower-visiting insects. Phytochemistry 72(13):1647–1654
16. Matignon D (1996) Stability results for fractional differential equations with applications to control processing. Comput Eng Syst Appl 2:963–968
17. Njagarah JBH, Tabi CB (2018) Spatial synchrony in fractional order metapopulation cholera transmission. Chaos Solitons Fractals 117:37–49 Dec

18. Oldham KB, Spanier J (1974) The fractional calculus theory and applications of differentiation and integration to arbitrary order. Academic Press, New York
19. Podlubny I (2001) Geometric and physical interpretation of fractional integration and fractional differentiation
20. Sánchez-Garduño F, Breña-Medina VF (2011) Searching for spatial patterns in a pollinator-plant-herbivore mathematical model. Bull Mathem Biol 73(5):1118–1153 May
21. Sánchez-Garduño F, Castellanos V, Quilantán I (2014) Dynamics of a nonlinear mathematical model for three interacting populations. Boletín de la Sociedad Matemática Mexicana 20(1):147–170 Apr
22. Sánchez Muñoz JM (2011) Génesis y desarrollo del Cálculo Fraccional. Pensamiento Matemático, pp 1–10
23. Sanjerehei M (2014) The economic value of bees as pollinators of crops in Iran. Ann Res Rev Biol 4(19):2957–2964 Jan
24. Soberon JM, Martinez Del Rio C (1981) The dynamics of a plant-pollinator interaction. J Theoret Biol 91(2):363–378 Jul
25. Soberón-Mainero J (2002) Ecología de Poblaciones, 3rd edn. Fondo de cultura económica, México
26. Sorensen JS, Dearing MD (2006) Efflux transporters as a novel herbivore Countermechanism to plant chemical defenses. J Chem Ecol 32(6):1181–1196

Chapter 4
Bayesian Inversion for a Fractional Order Harmonic Oscillator

V. R. Olvera-Reyes, M. P. Árciga-Alejandre, J. Sanchez-Ortiz, and Y. Peña-Pérez

4.1 Introduction

Differential equations of fractional order has been object of study of great interest in recent decades, this is due to its capacity to describe more accurately phenomena in many applied fields of knowledge; such as neurology, economics, physics, among others. For example, in physics several oscillators have been generalized using fractional operators [1, 2], such as the Caputo derivative or the fractional Lagrangian, applying solution methods based on the Grünwald-Letnikov approximation and the differential transform method [3–5]. In [6], the authors have introduced a fractional oscillator using a composition of Caputo and Riemann-Liouville fractional derivatives, approximating the solution by numerical algorithms and giving some examples. Also, in [7] a fractional friction term is added to the fractional oscillator and via Laplace transform an analytical solution is obtained.

In recent years, some authors have defined differential and integral operators of random order that have been used in problems governed by fractional differential equations, presenting examples where the order of the derivative is considered to be a simple random variable [8–11]. On the other hand, Bayesian statistics has been use to estimate the data involved in fractional differential equation models [12], as

V. R. Olvera-Reyes (✉) · M. P. Árciga-Alejandre · J. Sanchez-Ortiz · Y. Peña-Pérez
Universidad Autónoma de Guerrero (Autonomous University of Guerrero), Chilpancingo, Guerrero, México
e-mail: vanesa.or14@gmail.com

M. P. Árciga-Alejandre
e-mail: mparciga@uagro.mx

J. Sanchez-Ortiz
e-mail: jsanchez@uagro.mx

Y. Peña-Pérez
e-mail: ypenap88@gmail.com

J. J. Rosales García et al. (eds.), *Advances in Fractional Calculus*,
Modeling and Optimization in Science and Technologies 21,
https://doi.org/10.1007/978-3-031-83989-4_4

is the Lotka-Volterra equation [13], and some fractional diffusion equations, which are used to describe the behavior of cell migration [14].

In this work, we generalize the simple harmonic oscillator, which is a process that describes a periodic motion, considering the derivative order as a piecewise constant function. In each subinterval, where the order of the derivative is constant, using the semigroup property of the Riemann–Liouville fractional integral, we consider the associated integral equation instead of the fractional order differential equation. Using Gauss–Legendre quadrature method, we obtain the approximate solution, which allows observing the behavior of the oscillator, when the value of the derivative order and the spring constant varies. We generated synthetic data, and relying on the JAGS tool [15], the parameters involved are estimated.

4.2 Preliminaries

The concepts and properties necessary for the development of this work are presented below.

Definition 4.1 ([16]) Let $\alpha : [0, T] \to (1, 2]$ a piecewise constant function and $x \in L_1[0, T]$; the Riemann-Liouville fractional integral operator of time-variable order is defined by

$$I^{\alpha(t)}x(t) = \frac{1}{\Gamma(\alpha(t))} \int_0^t (t-\tau)^{\alpha(t)-1} x(\tau) d\tau.$$

In what follows, we will denote the derivative order using superscripts. For example, $D^2 = \frac{d^2}{dt^2}$.

Definition 4.2 ([16]) Let $\alpha : [0, T] \to (1, 2]$ a piecewise constant function and $x \in AC^2[0, T]$; the Caputo fractional differential operator of time-variable order is defined by

$${}^cD^{\alpha(t)}x(t) := \frac{1}{\Gamma(2-\alpha(t))} \int_0^t (t-\tau)^{1-\alpha(t)} D^2x(\tau) d\tau.$$

For constant α, the operators defined above are related as follows.

Theorem 4.1 ([17]) *Let $\alpha \geq 0$ constant, $n = \lceil \alpha \rceil$, and $x \in AC^2[0, T]$. Then*

$$I^{\alpha} \, {}^cD^{\alpha}x(t) = x(t) - \sum_{k=0}^{n-1} \frac{D^k x(0)}{k!} t^k.$$

Now, let's define an integral equation that we will find later in our work.

Definition 4.3 ([18]) The Fredholm linear integral equation of the second kind with constant limits of integration have a form

$$g(t) = x(t) - \lambda \int_a^b K(t,s)x(s)ds,$$

where $x(t)$ is the unknown function ($a \leq t \leq b$), $K(t,s)$ is the kernel of the integral equation, and $g(t)$ is a given function.

Finally, in the following theorem we give the numerical method, that we will use to approximate the solution.

Theorem 4.2 (Gauss–Legendre quadrature [19]) *Let the integral $\int_{t_i}^{t_{i+1}} f(t)dt$ over an arbitrary interval $[t_i, t_{i+1}]$, then it can be estimated by the Gaussian quadrature as follows*

$$\begin{aligned}\int_{t_i}^{t_{i+1}} f(t)dt &= \int_{-1}^{1} f\left(\frac{(t_{i+1}-t_i)\xi + (t_{i+1}+t_i)}{2}\right)\frac{t_{i+1}-t_i}{2}d\xi \\ &\approx \frac{t_{i+1}-t_i}{2}\sum_{l=1}^{n} f\left(\frac{(t_{i+1}-t_i)\xi_l + (t_{i+1}+t_i)}{2}\right)w_l,\end{aligned}$$

where

$$w_l = \int_{-1}^{1}\prod_{\substack{j=1\\ j\neq l}}^{n}\frac{\xi-\xi_j}{\xi_l-\xi_j}d\xi$$

are "weights" corresponding to the "nodes" (abscissa) $f(t_l)$, $(l = 1, ..., n)$, related to the roots of Legendre polynomials.

4.3 Fractional Harmonic Oscillator

In this section, we consider a boundary value problem for the fractional equation of the harmonic oscillator with variable order Caputo derivative,

$$^{c}D^{\alpha(t)}x(t) = -\lambda^2 x(t), \quad t \in [0,T], \quad \lambda \in \mathbb{R}^+ \tag{4.1}$$

$$x(0) = x_0, \quad x(T) = x_1, \quad x_0, x_1 \in \mathbb{R}^+,$$

where $\alpha : [0,T] \to (1,2]$ is a piecewise constant function. Given the partition $0 = t_0 < t_1 < ... < t_n = T$, of the time interval $[0,T]$, assume that the derivative order is constant in each of the subintervals $[t_i, t_{i+1}]$, $i = 0, ..., n-1$. For $t \in [t_i, t_{i+1}]$, the problem (4.1) is reformulated in the following way

$$^{c}D^{\alpha_i}x(t) = -\lambda^2 x(t), \quad t \in [t_i, t_{i+1}], \quad \lambda \in \mathbb{R}^+, \quad \alpha_i \in (1,2], \tag{4.2}$$

$$x(t_i) = x_i, \quad x(t_{i+1}) = x_{i+1}, \quad x_i, x_{i+1} \in \mathbb{R}^+.$$

Note that, the final boundary condition in $[t_{i-1}, t_i]$ corresponds to the initial boundary condition in $[t_i, t_{i+1}]$.

Applying the Riemann-Liouville fractional integral of order α_i on both sides of the differential equation of the system (4.2), by the Definition 4.1 and the Theorem 4.1, we obtain

$$x(t) = c_0 + c_1(t - t_i) - \lambda^2 \frac{1}{\Gamma(\alpha_i)} \int_{t_i}^{t} (t - s)^{\alpha_i - 1} x(s) ds, \tag{4.3}$$

where $t \in [t_i, t_{i+1}]$. Evaluating $x(t)$ in t_i and t_{i+1}, we have $c_0 = x_i$ and

$$c_1 = \frac{x_{i+1} - x_i}{t_{i+1} - t_i} + \frac{\lambda^2}{t_{i+1} - t_i} \frac{1}{\Gamma(\alpha_i)} \int_{t_i}^{t_{i+1}} (t_{i+1} - s)^{\alpha_i - 1} x(s) ds.$$

Then substituting c_0 and c_1 in (4.3), we get

$$\begin{aligned} x(t) = {} & x_i + \frac{x_{i+1} - x_i}{t_{i+1} - t_i}(t - t_i) + \lambda^2 \frac{t - t_i}{t_{i+1} - t_i} \frac{1}{\Gamma(\alpha_i)} \int_{t_i}^{t_{i+1}} (t_{i+1} - s)^{\alpha_i - 1} x(s) ds \\ & - \lambda^2 \frac{1}{\Gamma(\alpha_i)} \int_{t_i}^{t} (t - s)^{\alpha_i - 1} x(s) ds. \end{aligned}$$

Expressing the integral of the third summand of the right hand side of the equation as the sum of two integrals, from the fact that $(t_i, t_{i+1}) = (t_i, t) \cup (t, t_{i+1})$, we obtain

$$\begin{aligned} x(t) = {} & x_i + \frac{x_{i+1} - x_i}{t_{i+1} - t_i}(t - t_i) + \lambda^2 \frac{t - t_i}{t_{i+1} - t_i} \frac{1}{\Gamma(\alpha_i)} \int_{t}^{t_{i+1}} (t_{i+1} - s)^{\alpha_i - 1} x(s) ds \\ & + \lambda^2 \frac{1}{\Gamma(\alpha_i)} \int_{t_i}^{t} \left[\frac{t - t_i}{t_{i+1} - t_i} (t_{i+1} - s)^{\alpha_i - 1} - (t - s)^{\alpha_i - 1} \right] x(s) ds, \end{aligned}$$

therefore we can express $x(t)$ as a Fredholm equation of the second type

$$x(t) = g(t) - \lambda^2 \int_{t_i}^{t_{i+1}} G(t, s) x(s) ds, \tag{4.4}$$

where

$$g(t) = x_i + (x_{i+1} - x_i) \frac{t - t_i}{t_{i+1} - t_i},$$

and

$$G(t,s)=\begin{cases}\frac{1}{\Gamma(\alpha_i)}\left[\frac{t-t_i}{t_{i+1}-t_i}(t_{i+1}-s)^{\alpha_i-1}-(t-s)^{\alpha_i-1}\right], & t_i\le s\le t,\\ \frac{1}{\Gamma(\alpha_i)}\left[\frac{t-t_i}{t_{i+1}-t_i}(t_{i+1}-s)^{\alpha_i-1}\right], & t\le s\le t_{i+1}.\end{cases}$$

4.3.1 Numerical Solution

The Gauss-Legendre quadrature method is used to approximate the solution of the integral equation for $x(t)$. Using Theorem 4.2, we can express (4.4) as follows

$$x(t)=g(t)-\lambda^2\sum_{j=1}^{N}G(t,s_j)x(s_j)w_j.$$

Then, evaluating $x(t)$ at the same nodes of the quadrature rule, we obtain

$$x(t_k)=g(t_k)-\lambda^2\sum_{j=1}^{N}G(t_k,s_j)x(s_j)w_j, \tag{4.5}$$

with

$$t_k=s_k=\left(\frac{(t_{i+1}-t_i)\xi_k+(t_{i+1}+t_i)}{2}\right).$$

Now, we express (4.5) in matrix form

$$\mathbf{x}=\mathbf{g}-\lambda^2G\mathbf{x},$$

where

$$\mathbf{x}=\begin{pmatrix}x(t_1)\\ \vdots\\ x(t_N)\end{pmatrix},\quad \mathbf{g}=\begin{pmatrix}g(t_1)\\ \vdots\\ g(t_N)\end{pmatrix},$$

$$G=\begin{pmatrix}G(t_1,s_1)w_1 & G(t_1,s_2)w_2 & \dots & G(t_1,s_N)w_N\\ G(t_2,s_1)w_1 & G(t_2,s_2)w_2 & \dots & G(t_2,s_N)w_N\\ \vdots & \vdots & \ddots & \vdots\\ G(t_N,s_1)w_1 & G(t_N,s_2)w_2 & \dots & G(t_N,s_N)w_N\end{pmatrix}.$$

Note that, if $\frac{1}{\lambda^2}$ is not an eigenvalue of G, then $I+\lambda^2G$ is invertible and in this case

$$\mathbf{x}=[I+\lambda^2G]^{-1}\mathbf{g},$$

which is a discrete solution of the problem (4.2). Using the Gaussian quadrature package in the R/RStudio software, we find the weights, from which we obtain approximated values of $x(t)$. For example, for $x(t_0) = 50$ and $x(t_n) = 180$, we illustrate the behavior of the fractional order harmonic oscillator for different values of α and λ (Figs. 4.1, 4.2, 4.3, 4.4, 4.5 and 4.6).

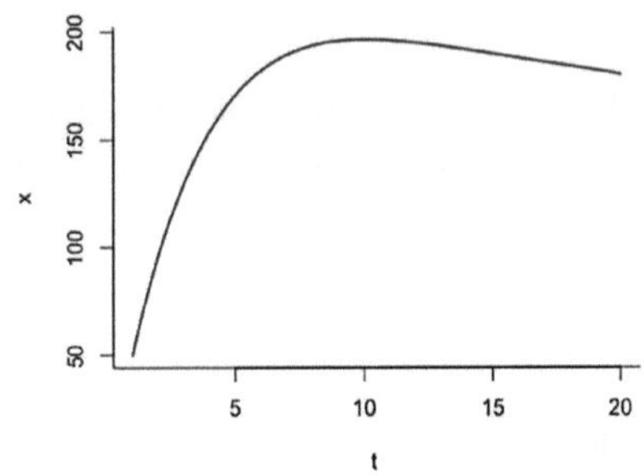

Fig. 4.1 Solution behavior for $\alpha = 1.1, \lambda = 0.5$

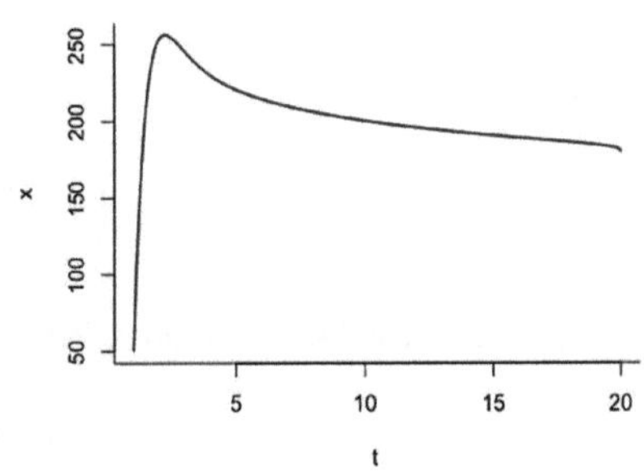

Fig. 4.2 Solution behavior for $\alpha = 1.1, \lambda = 1.5$

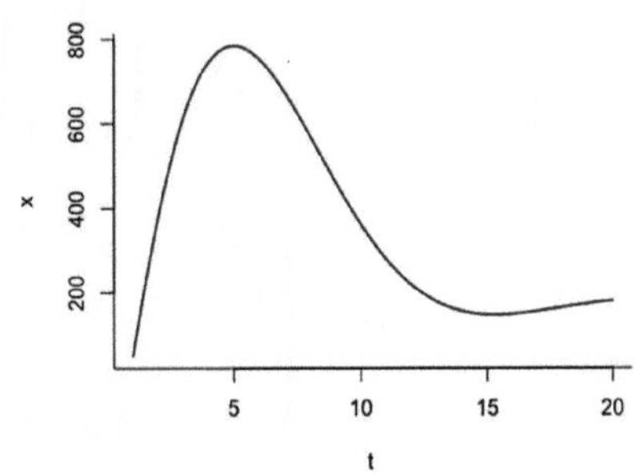

Fig. 4.3 Solution behavior for $\alpha = 1.5, \lambda = 0.5$

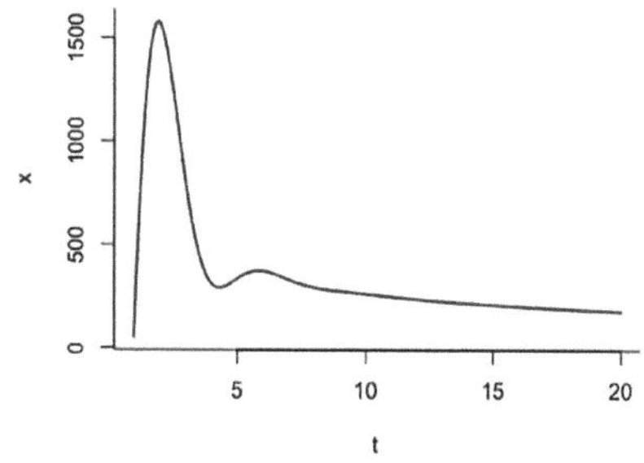

Fig. 4.4 Solution behavior for $\alpha = 1.5$, $\lambda = 1.5$

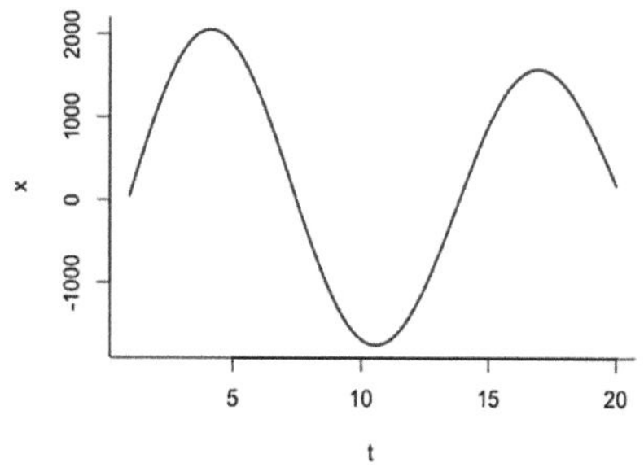

Fig. 4.5 Solution behavior for $\alpha = 1.95$, $\lambda = 0.5$

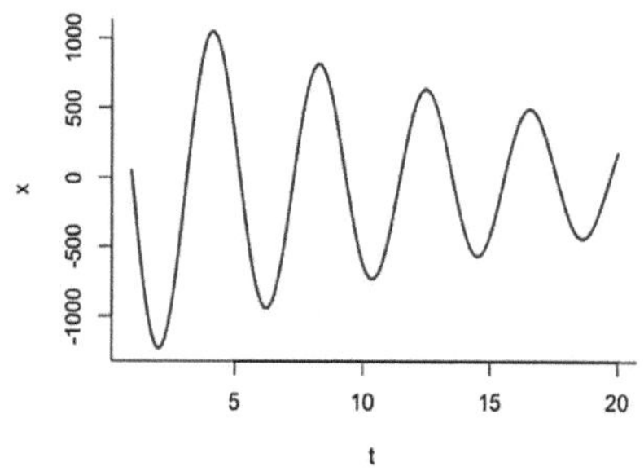

Fig. 4.6 Solution behavior for $\alpha = 1.95$, $\lambda = 1.5$

Note that, α controls the "number of cycles" of the oscillator and λ the wavelength, increasing the wave amplitude as both values increment.

4.4 Inverse Problem

To analyze the performance of parameter estimation, we first generate synthetic data, with the following algorithm.

1. Using the imputation method, we generate the data for each time instant, with the same initial and final values used in the analysis of the direct problem and fixed values for the order of the derivative and the constant of the oscillator equation.
2. A disturbance is performed by adding to the random data an error, which is distributed as a normal with constant variance and mean zero.

Now, using the synthetic data generated previously, the parameters $\lambda \in [0, 2]$ and $\alpha \in [1, 2]$ are estimated via Bayesian Inference with the following prior distributions

$$\lambda \sim N(0.5, 1)\mathbb{I}(0, 2),$$

$$\alpha \sim N(1.5, 10)\mathbb{I}(1, 2),$$

where $N(\mu, \sigma)$ is the normal distribution with mean μ and variance σ, and $\mathbb{I}$ the indicator function.

Now, an example is given to illustrate the method: first, using $\alpha = 1.1$ and $\lambda = 0.5$ we generate synthetic data. Then, using the above prior distributions we obtain posterior distributions through Bayesian Inference [20, 21], from which the means are plotted below (Figs. 4.7, 4.8 and 4.9).

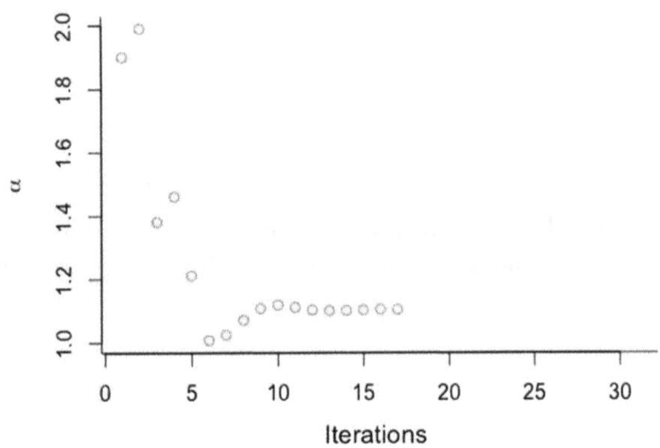

Fig. 4.7 α-value estimation for each iteration

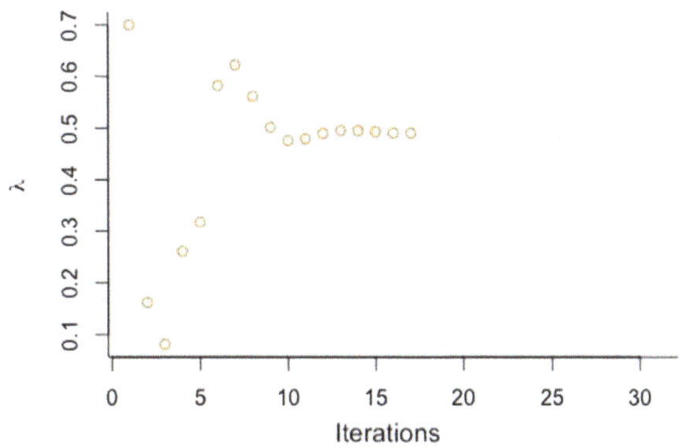

Fig. 4.8 λ-value estimation for each iteration

Finally, using the mean values obtained in the estimation for α and λ, we plot the solution (4.2) and the synthetic data

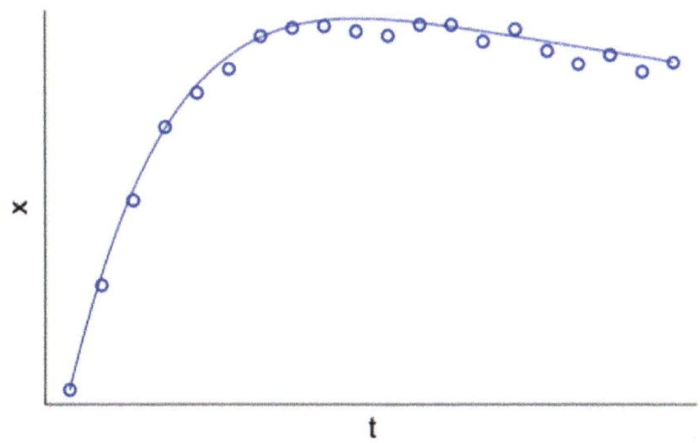

Fig. 4.9 Synthetic data for $\alpha = 1.1$, $\lambda = 0.5$ and estimated solution

4.4.1 Simulation Study

In order to give evidence of stability for the Bayesian estimators, 50 different sets of synthetic data were generated for constant values of λ and α, estimating the parameter values for each of them. Once the results of these estimations were obtained, the mean squared error (MSE) was calculated, thus observing numerically the effectiveness of the process.

The results are presented in the following table.

Table 4.1 Mean, variance, bias and ECM of α-value and λ-value estimates from 50 synthetic data sets

Parameters	Mean	Variance	Bias	MSE
$\alpha = 1.1$	1.10783864	0.001298086	0.00783864	0.00135953
$\lambda = 0.5$	0.485801235	0.00223666	−0.01419877	0.002438265
$\alpha = 1.1$	1.094762	1.220111e-05	−0.00523756	3.963315e-05
$\lambda = 1.5$	1.519864	0.003531973	0.01986354	0.003926533
$\alpha = 1.25$	1.246989	0.0008122031	−0.0030114	0.0008212717
$\lambda = 0.5$	0.5058268	0.001323198	0.00582684	0.00135715
$\alpha = 1.25$	1.242961	0.002018507	−0.00703926	0.002068058
$\lambda = 1.5$	1.738465	0.2055451	0.2384646	0.2624105
$\alpha = 1.5$	1.489861	0.03159698	−0.01013862	0.03169977
$\lambda = 0.5$	0.3995827	0.1409535	−0.1004173	0.1510372
$\alpha = 1.5$	1.461491	0.01485166	−0.03850866	0.01633458
$\lambda = 1.5$	1.570003	0.5266609	0.07000266	0.5315613
$\alpha = 1.75$	1.543277	0.01320598	−0.2067234	0.05594056
$\lambda = 0.5$	0.5649965	0.1618893	0.06499652	0.1661138
$\alpha = 1.75$	1.731618	0.003679848	−0.01838164	0.004017733
$\lambda = 1.5$	0.2960688	0.01582605	−1.203931	1.465276
$\alpha = 1.95$	1.564797	0.006787082	−0.3852032	0.1551686
$\lambda = 0.5$	0.7826447	0.0236703	0.2826447	0.1035583
$\alpha = 1.95$	1.70092	0.001059763	−0.2490799	0.06310054
$\lambda = 1.5$	0.1875029	0.009175317	−1.312497	1.731824

As shown in Table 4.1, the bias and MSE were very close to zero, indicating stability of the Bayesian parameter estimates.

Acknowledgements The authors gratefully acknowledge the financial support of the Postgraduate Study Fellowship by CONAHCYT, Mexico, (grant number: 1106351).

Competing Interests The authors have no conflicts of interest to declare that are relevant to the content of this chapter.

References

1. Duan JS (2013) The periodic solution of fractional oscillation equation with periodic input. Adv Mathem Phys 2013:1–6
2. Gómez-Aguilar JF, Rosales-García JJ, Bernal-Alvarado JJ, Córdoba-Fraga T, Guzmán-Cabrera R (2012) Fractional mechanical oscillators. Revista Mexicana de Física 58(4):348–352
3. Al-rabtah A, Ertürk VS, Momani S (2010) Solutions of a fractional oscillator by using differential transform method. Comput Mathem Appl 59(3):1356–1362

4. Baleanu D, Asad JH, Petras I (2014) Fractional bateman-feshbach-tikochinsky oscillator. Commun Theoret Phys 61(2):221–225
5. Baleanu D, Petras I, Asad JH, Velasco MP (2011) Fractional pais-uhlenbeck oscillator. Int J Theoret Phys 51(4):1253–1258
6. Blaszczyk T, Ciesielski M (2015) Fractional oscillator equation: analytical solution and algorithm for its approximate computation. J Vib Control 22(8):2045–2052
7. Duan JS, Zhang JY, Qiu X (2022) Exact solutions of fractional order oscillation equation with two fractional derivative terms. J Nonlinear Mathem Phys
8. Lorenzo CF, Hartley TT (2002) Variable order and distributed order fractional operators. Nonlinear Dyn 29:57–98
9. Valério D, da Costa JS (2011) Variable-order fractional derivatives and their numerical approximations. Signal Process 91(3):470–483
10. Benkerrouche A, Souid MS, Sitthithakerngkiet K, Hakem A (2021) Implicit nonlinear fractional differential equations of variable order. Boundary Value Probl 2021:1–16
11. Lopez-Cresencio OU, Ariza-Hernandez FJ, Sanchez-Ortiz J, Arciga-Alejandre MP (2022) A boundary value problem for a random-order fractional differential equation. Results Appl Mathem 16:100328
12. Xiong-Bin Y, Wei T (2023) A fractional order and time-dependent coefficient in a time-fractional diffusion wave equation. J Comput Appl Mathem 424:112995
13. Ariza-Hernandez FJ, Martin-Alvarez LM, Arciga-Alejandre MP, Sanchez-Ortiz J (2021) Bayesian inversion for a fractional Lotka-Volterra model: An application of Canadian lynx vs. snowshoe hares. Chaos Solitions Fractals 151: 111278
14. Ariza-Hernandez FJ, Najera-Tinoco JC, Arciga-Alejandre MP, Castaneda-Saucedo E, Sanchez-Ortiz J (2022) Bayesian inverse problem for a fractional diffusion model of cell migration. Mathem Biosci Eng 21(4):5826–5837
15. Plummer M (2024) rjags: Bayesian graphical models using MCMC. R package version 4–16
16. Samko S (2013) Fractional integration and differentiation of variable order: an overview. Nonlinear Dyn 71:653–662
17. Diethelm K (2010) The analysis of fractional differential equations. Springer, Berlin, Heidelberg
18. Polyanin AD, Manzhirow AV (2008) Handbook of integral equations. Chapman and Hall/CRC
19. Burden RL, Douglas JF, Burden AM (2016) Numerical analysis. BROOKS/COLE, CENGAGE Learning
20. Box GE, Tiao GC (2011) Bayesian inference in statistical analysis. Wiley
21. Congdon P (2006) Bayesian statistical modelling. Wiley

Chapter 5
Dirichlet Problem for a Linear Time-Fractional KdV Equation on a Half-Line

J. L. Escobar-Figueroa, M. P. Árciga-Alejandre, J. Sanchez-Ortiz, and Y. Peña-Pérez

5.1 Introduction

The Korteweg-de Vries (KdV) equation

$$u_t + u_x + uu_x + u_{xxx} = 0, \tag{5.1}$$

is one of the most studied nonlinear partial differential equations (PDEs) [1–3]. It arises in the study of ion-acoustic waves in plasmas, long waves in shallow water, and in general, describes the slow evolution of long waves in dispersive media [4].

The KdV equation plays an important role in the description of solitary waves, lattice dynamics, plasma physics, stratified internal waves, collision-free hydromagnetic waves, etc., see for instance [3, 5]. There are various methods available for solving boundary value problem related to the KdV equation. These include the finite difference method, variational iteration method, homotopy perturbation method, scattering method for exact solution, Backlund transformations, etc., see [3, 6, 7] and the references therein.

In the 90s, Athanassios Fokas [8] introduced the unified transform method to find integral representations of the solution of several equations. This method, so-called

J. L. Escobar-Figueroa (✉) · M. P. Árciga-Alejandre · J. Sanchez-Ortiz · Y. Peña-Pérez
Universidad Autónoma de Guerrero, Chilpancingo, Guerrero, México
e-mail: jlef.mate@gmail.com

M. P. Árciga-Alejandre
e-mail: mparciga@uagro.mx

J. Sanchez-Ortiz
e-mail: jsanchez@uagro.mx

Y. Peña-Pérez
e-mail: ypenap88@gmail.com

J. J. Rosales García et al. (eds.), *Advances in Fractional Calculus*,
Modeling and Optimization in Science and Technologies 21,
https://doi.org/10.1007/978-3-031-83989-4_5

Fokas method [9, 10], has been used successfully to solve different types of initial boundary-value problems for linear KdV equation, e.g., [11–14].

In recent years, numerical methods have been used to approximate the solution of the KdV equation with fractional derivatives [15–18]; moreover, study of invariant subspaces, asymptotic analysis and exact solutions of the fractional KdV equation are studied in [19–21]. In this work, we study a linear time fractional KdV equation, where the fractional derivative considered is in the Riemann-Liouville sense.

The organization of the paper is the following. We collect in Sect. 5.2 the definitions of Laplace transform and Caputo fractional derivative, some notations which are standard for the fractional differential equations theory and the problem is stated. Section 5.3 is devoted to the proof of the main result.

5.2 Preliminaries

In this section, we discuss the terminology and establish the notation and properties used in the research.

Definition 5.1 The Laplace transform with respect to space and time is defined as

$$\widehat{u}(k,\eta) = \int_0^\infty \int_0^\infty e^{-ikx-i\eta t} u(x,t)dxdt,$$

where $x, t \geq 0$, $\Im m\, k < 0$, $\Im m\, \eta < 0$ and its inverse transform by

$$u(x,t) = \frac{1}{(2\pi)^2} \int_{\mathbb{R}} \int_{\mathbb{R}} e^{ikx+i\eta t} \widehat{u}(k,\eta) dk d\eta.$$

In particular, the Laplace transform with respect only to space is $\widehat{u}(k,t) = \int_0^\infty e^{-ikx} u(x,t)dx$ and respect only to time is $\widehat{u}(x,\eta) = \int_0^\infty e^{-i\eta t} u(x,t)dt$.

For the convenience of the reader we repeat the following lemma from [22] without proofs, thus making our exposition self-contained.

Lemma 5.1 *The Laplace transform of the function* $\Phi(t) = \frac{t^{n-(1+\alpha)}}{\Gamma(n-\alpha)}$ *is*

$$\widehat{\Phi}(\eta) = (i\eta)^{\alpha-n}, \quad n \in \mathbb{N},$$

where Γ *is the gamma function.*

Definition 5.2 Let $u(x,t)$ an absolutely continuous function with respect to t. The Caputo fractional derivative of order $\alpha \in (0, 1]$, is defined by

$${}^C D_t^\alpha u = \frac{1}{\Gamma(1-\alpha)} \int_0^t \frac{u_\tau(x,\tau)}{(t-\tau)^\alpha} d\tau.$$

Let us note that the above integral can be written through a convolution

$$ {}^{C}D_t^{\alpha}u = \left[\frac{t^{-\alpha}}{\Gamma(1-\alpha)} * u_t\right](x,t). \tag{5.2}$$

Definition 5.3 The two-parametric Mittag-Leffler function is define by

$$E_{\alpha,\beta}[z] = \sum_{k=0}^{\infty} \frac{z^k}{\Gamma(\alpha k+\beta)}, \tag{5.3}$$

where $\alpha > 0$ and $\beta \in \mathbb{C}$.

An immediate property of the Mittag-leffler function

$$\left(\widehat{t^{\beta-1}E_{\alpha,\beta}[\lambda t^{\alpha}]}\right)(s) = \frac{s^{\alpha-\beta}}{s^{\alpha}-\lambda}, \tag{5.4}$$

for $\Re e(\beta) > 0$, $\lambda \in \mathbb{C}$ and $|\lambda s^{-\alpha}| < 1$ (see [23]).

In the next section, we study the following initial-boundary value problem for a fractional linear KdV equation

$$\begin{aligned} {}^{C}D_t^{\alpha}u(x,t) &= -u_{xxx}(x,t), \quad x>0, \quad t>0, \\ u(x,0) &= u_0(x), \quad x>0, \\ u(0,t) &= g_0(t), \quad t>0, \end{aligned} \tag{5.5}$$

where $\alpha \in (0,1]$.

5.3 Dirichlet Problem for a Fractional Linear KdV Equation

The following theorem, our main result, states a generalization of Problem 3.2 in [11]. Its proof will take essentially the remaining of the paper.

Theorem 5.1 *Let $u_0(x) \in \mathbf{L}^1(\mathbb{R})$ and $g_0(t) \in \mathbf{C}(\mathbb{R})$, assume that there is some function $u(x,t)$ satisfying (5.5), then $u(x,t)$ has the following integral representation*

$$\begin{aligned} u(x,t) &= \tfrac{1}{2\pi}\int_{\mathbb{R}} e^{ikx}E_{\alpha}\left(ik^3t^{\alpha}\right)\widehat{u_0}(k)dk - \tfrac{3}{2\pi}\int_{\partial D^+} k^2e^{ikx}G_0(ik^3,t)dk \\ &+ \tfrac{1}{2\pi}\int_{\partial D^+} e^{ikx}E_{\alpha}\left(ik^3t^{\alpha}\right)\left[\nu\widehat{u_0}(k\nu)+\nu^2\widehat{u_0}(k\nu^2)\right]dk, \end{aligned}$$

where

$$G_0(ik^3,t) = \int_0^t (t-\tau)^{\alpha-1}E_{\alpha,\alpha}[ik^3(t-\tau)^{\alpha}]g_0(\tau)d\tau.$$

Proof First, applying the Laplace transform to both sides of (5.5) with respect to space and time

$$\int_0^{\infty} e^{-ikx}\left[\int_0^{\infty} \frac{t^{-\alpha}}{\Gamma(1-\alpha)} e^{-i\eta t}dt \cdot \int_0^{\infty} e^{-i\eta t} u_t(x,t)dt\right]dx =$$

$$-\int_0^{\infty} e^{-i\eta t}\left[\int_0^{\infty} e^{-ikx} u_{xxx}(x,t)dx\right]dt,$$

using the Lemma 5.1 and Eq. (5.2), we arrive to

$$\widehat{u}(k,\eta) = \frac{1}{(i\eta)^{\alpha} + (ik)^3}\left[(i\eta)^{\alpha-1}\widehat{u}(k,0) + k^2\widehat{u}(0,\eta)\right.$$
$$\left. -ik\widehat{u}_x(0,\eta) - \widehat{u}_{xx}(0,\eta)\right]. \tag{5.6}$$

Now, taking the inverse Laplace transform with respect to time in (5.6), we obtain

$$\widehat{u}(k,t) = \tfrac{1}{2\pi}\int_{\mathbb{R}} \frac{(i\eta)^{\alpha-1}e^{i\eta t}}{(i\eta)^{\alpha} + (ik)^3}\widehat{u}(k,0)d\eta + \tfrac{k^2}{2\pi}\int_{\mathbb{R}} \frac{e^{i\eta t}}{(i\eta)^{\alpha} + (ik)^3}\widehat{u}(0,\eta)d\eta$$
$$-\tfrac{ik}{2\pi}\int_{\mathbb{R}} \frac{e^{i\eta t}}{(i\eta)^{\alpha} + (ik)^3}\widehat{u}_x(0,\eta)d\eta - \tfrac{1}{2\pi}\int_{\mathbb{R}} \frac{e^{i\eta t}}{(i\eta)^{\alpha} + (ik)^3}\widehat{u}_{xx}(0,\eta)d\eta.$$

By (5.4) and the inverse Laplace transform of a product, we arrive at the expression

$$\widehat{u}(k,t) = \widehat{u}(k,0)E_{\alpha}\left(ik^3t^{\alpha}\right) - k^2G_0(ik^3,t) + ikG_1(ik^3,t) + G_2(ik^3,t), \tag{5.7}$$

where

$$G_j(ik^3,t) = \int_0^t (t-\tau)^{\alpha-1}E_{\alpha,\alpha}\left(ik^3(t-\tau)^{\alpha}\right)\partial_x^j g_0(\tau)\,d\tau, \quad j = 0,1,2.$$

The Eq. (5.7), called global relation is valid for $\Im m\, k < 0$. Now, we must analyze the numbers $k \in \mathbb{C}$ that satisfy $\Im m(-ik^3) < 0$, obtain the regions

$$D^+ = \left\{k : \arg k \in (\pi/3, \frac{2\pi}{3})\right\},$$

and

$$D_1^- = \left\{k : \arg k \in (\pi, \frac{4\pi}{3})\right\}, \quad D_2^- = \left\{k : \arg k \in (\frac{5\pi}{3}, 2\pi)\right\}.$$

Then, applying the inverse Laplace transform with respect to t in (5.7) we arrive at the relation

$$
\begin{aligned}
u(x,t) = {} & \tfrac{1}{2\pi} \int_{\mathbb{R}} e^{ikx} E_\alpha\left(ik^3 t^\alpha\right) \widehat{u}_0(k) dk \\
& + \tfrac{1}{2\pi} \int_{\mathbb{R}} e^{ikx} \left[-k^2 G_0(ik^3,t) + ik G_1(ik^3,t) + G_2(ik^3,t)\right] dk.
\end{aligned}
$$

Since the functions $G_j(ik^3, t)$ are analytic for $k \in \mathbb{C}$, we can deform the second integral in the upper complex half-plane to the boundary of the region D^+ in the above equation

$$
\begin{aligned}
u(x,t) = {} & \tfrac{1}{2\pi} \int_{\mathbb{R}} e^{ikx} E_\alpha\left(ik^3 t^\alpha\right) \widehat{u}_0(k) dk - \tfrac{1}{2\pi} \int_{\partial D^+} k^2 e^{ikx} G_0(ik^3,t) dk \\
& + \tfrac{1}{2\pi} \int_{\partial D^+} e^{ikx} \left[ik G_1(ik^3,t) + G_2(ik^3,t)\right] dk. \qquad (5.8)
\end{aligned}
$$

On the other hand, the expression $u(x, t)$ of the Eq. (5.8) does not represent the solution of the problem (5.5) because unknown boundary terms appear. Then, we use the cube roots of the unit $\nu = e^{\frac{2\pi i}{3}}$, $\nu^2 = e^{\frac{4\pi i}{3}}$, and we note that for $k \in D^+$, $\nu k \in D_1^-$ and $\nu^2 k \in D_2^-$. Applying these transformations to the Eq. (5.7) for $k \in D^+$ we obtain the system

$$
\widehat{u}(k\nu,t) = E_\alpha\left(ik^3 t^\alpha\right) \widehat{u}_0(k\nu) - k^2\nu^2\, G_0(ik^3,t) + ik\nu\, G_1(ik^3,t) + G_2(ik^3,t)
$$

$$
\widehat{u}(k\nu^2,t) = E_\alpha\left(ik^3 t^\alpha\right) \widehat{u}_0(k\nu^2) - k^2\nu\, G_0(ik^3,t) + ik\nu^2\, G_1(ik^3,t) + G_2(ik^3,t).
$$

Solving the above equations for G_1, G_2 and $k \in D^+$, we find

$$
\begin{aligned}
G_1(ik^3,t) = {} & \frac{1}{ik(\nu-\nu^2)} \left[\widehat{u}(k\nu,t) - E_\alpha\left(ik^3 t^\alpha\right) \widehat{u}_0(k\nu) + k^2\nu^2 G_0(ik^3,t)\right. \\
& \left. -\widehat{u}(k\nu^2,t) + E_\alpha\left(ik^3 t^\alpha\right) \widehat{u}_0(k\nu^2) - k^2\nu G_0(ik^3,t)\right], \qquad (5.9)
\end{aligned}
$$

$$
\begin{aligned}
G_2(ik^3,t) = {} & \frac{1}{\nu-\nu^2} \left[-\nu^2 \widehat{u}(k\nu,t) + \nu^2 E_\alpha\left(ik^3 t^\alpha\right) \widehat{u}_0(k\nu) - k^2\nu G_0(ik^3,t)\right. \\
& \left. +\nu\widehat{u}(k\nu^2,t) - \nu E_\alpha\left(ik^3 t^\alpha\right) \widehat{u}_0(k\nu^2) + k^2\nu^2 G_0(ik^3,t)\right]. \qquad (5.10)
\end{aligned}
$$

Note that, the third term on the right-hand side of Eq. (5.8) contains the expression

$$
\begin{aligned}
ikG_1(ik^3,t) + G_2(ik^3,t) = {} & \nu E_\alpha(ik^3 t^\alpha) \widehat{u}_0(\nu k) + \nu^2 E_\alpha(ik^3 t^\alpha) \widehat{u}_0(\nu^2 k) \\
& -2k^2 G_0(ik^3,t) - \nu\widehat{u}(\nu k,t) - \nu^2 \widehat{u}(\nu^2 k,t),
\end{aligned}
$$

and substituting in (5.8) we get

$$\begin{aligned} u(x,t) = \quad & \tfrac{1}{2\pi}\int_{\mathbb{R}} e^{ikx} E_\alpha\left(ik^3 t^\alpha\right)\widehat{u}_0(k)dk - \tfrac{1}{2\pi}\int_{\partial D^+} k^2 e^{ikx} G_0(ik^3,t)dk \\ & + \tfrac{1}{2\pi}\int_{\partial D^+} e^{ikx}\left[\nu E_\alpha(ik^3t^\alpha)\widehat{u}_0(\nu k) + \nu^2 E_\alpha(ik^3t^\alpha)\widehat{u}_0(\nu^2 k)\right. \\ & \left. - 2k^2 G_0(ik^3,t) - \nu\widehat{u}(\nu k,t) - \nu^2\widehat{u}(\nu^2 k,t)\right]dk. \end{aligned} \tag{5.11}$$

Finally, Jordan's Lemma and Cauchy's theorem applied to the path of integration shown in the following figure

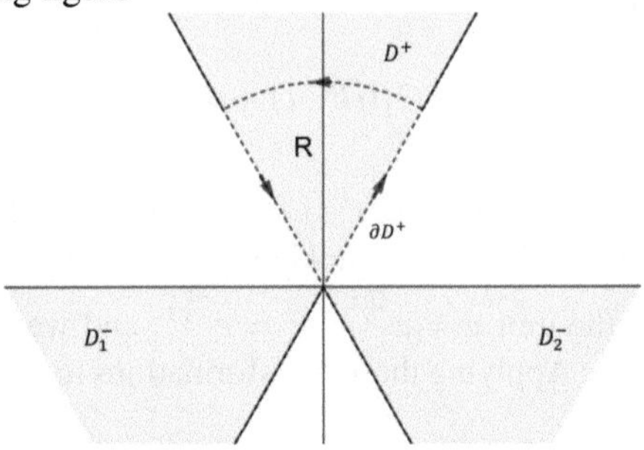

assures that the integrals

$$\int_{\partial D^+} e^{ikx}\widehat{u}(\nu k,t)dk, \qquad \int_{\partial D^+} e^{ikx}\widehat{u}(\nu^2 k,t)dk$$

do not contribute to the solution when $R \to \infty$, we obtain the desired result. □

Acknowledgements Jorge L. Escobar Figueroa gratefully acknowledges the financial support of the Postgraduate Study Fellowship of the Consejo Nacional de Humanidades, Ciencias y Tenologías (CONAHCYT) (Grant Number: 1106333).

Competing Interests The authors have no conflicts of interest to declare that are relevant to the content of this chapter.

References

1. Hirota R (1971) Exact solution of the Korteweg-de Vries equation for multiple collisions of solitons. Phys Rev Lett 27(18):1192–1194
2. Miura RM, Gardner CS, Kruskal MD (1971) Korteweg-de Vries equation and generalizations. II. Existence of conservation laws and constants of motion. J Mathem Phys 9(8):1204–1209
3. Miura RM (1976) The Korteweg-deVries equation: a survey of results. SIAM Rev 18(3):412–459
4. Ablowitz MJ, Segur H (1976) Solitons and the inverse scattering transform. Volume 4 of SIAM Studies in Applied Mathematic
5. Fung MK (1997) KdV Equation as an Euler-Poincare' equation. Chin J Phys 35(6):789

6. Appadu AR, Kelil AS (2021) Comparison of modified ADM and classical finite difference method for some third-order and fifth-order KdV equations. Demonstr Mathem 54:377–409
7. Kelil AS, Appadu AR (2022) On the numerical solution of 1D and 2D KdV equations using variational homotopy perturbation and finite difference methods. Mathematics 10:4443
8. Fokas AS (1997) A unified transform method for solving linear and certain nonlinear PDEs. Proc Royal Soc London Series A Mathem Phys Eng Sci 453(1962):1411–1443
9. Fokas AS, Pelloni B (1998) The solution of certain initial boundary-value problems for the linearized Korteweg-deVries equation. Proc Royal Soc London Series A Mathem Phys Eng Sci 454(1970):645–657
10. Fokas AS, Pelloni B (Eds) (2014) Unified transform for boundary value problems: applications and advances. Society for industrial and applied mathematics
11. Deconinck B, Trogdon T, Vasan V (2014) The method of Fokas for solving linear partial differential equations. SIAM Rev 56(1):159–186
12. Deconinck B, Sheils NE, Smith DA (2016) The linear KdV equation with an interface. Commun Mathem Phys 347:489–509
13. Chatziafratis A, Kamvissis S, Stratis IG (2023) Boundary behavior of the solution to the linear Korteweg-De Vries equation on the half line. Studies Appl Mathem 150(2):339–379
14. Chatziafratis A, Grafakos L, Kamvissis S (2024) Long-range instability of linear evolution PDE on semi-bounded domains via the Fokas method. Dyn PDE 21(2):97–169
15. El-Wakil SA, Abulwafa EM, Zahran MA, Mahmoud AA (2011) Time-fractional KdV equation: formulation and solution using variational methods. Nonlinear Dyn 65:55–63
16. Ray SS (2013) Soliton solutions for time fractional coupled modified KdV equations using new coupled fractional reduced differential transform method. J Mathem Chem 51(8):2214–2229
17. Yokus A (2018) Comparison of Caputo and conformable derivatives for time-fractional Korteweg-de Vries equation via the finite difference method. Int J Modern Phys B 32(29):1850365
18. Appadu AR, Kelil AS (2023) Some finite difference methods for solving linear fractional KdV equation. Front Appl Mathem Stat 9:1261270
19. Sahadevan R, Bakkyaraj T (2015) Invariant subspace method and exact solutions of certain nonlinear time fractional partial differential equations. Fract Calculus Appl Anal 18:146–162
20. Hayashi N, Naumkin PI, Sánchez-Suárez I (2022) Large time asymptotics for the fractional modified Korteweg-de Vries equation with $\alpha \in (2, 4)$. Partial Differ Equ Appl 3(6):76
21. Carreño-Bolaños R, Naumkin PI (2023) Large time asymptotics for the fractional modified Korteweg-de Vries equation of order $\alpha \in [4, 5)$. J Pseudo-Differe Oper Appl 14(3):42
22. Diethelm K, Ford NJ (2010) The analysis of fractional differential equations. Lecture notes in mathematics
23. Gorenflo R, Kilbas AA, Mainardi F, Rogosin SV (2020) Mittag-Leffler functions, related topics and applications. Springer, Berlin, p 540

Chapter 6
Analysis of Fractional-Order Neuronal Hysteresis Loops

L. R. González-Ramírez

6.1 Introduction

Hysteresis is characterized by a non-linear operator consisting of multiple stable steady states. It describes a phenomenon where an external input causes a transition between these states. This transition is highly dependent on the previous input's value and, as a consequence, hysteresis is conceived as a memory-dependent phenomenon. Hysteresis is encountered in various biological systems [1] (e.g., lung function and protein-DNA interactions) and physical systems [2] (e.g., magnet materials and thermostats).

Neuronal activity is intrinsically governed by distinct non-linear processes at different spatial and temporal scales. For example, there is ample evidence of hysteresis in visual and auditory perception tasks [3–6]. Research has shown that the hysteresis effects observed in these tasks are not due to physical movements of body parts (e.g., eyes) but rather to the underlying cortical mechanisms [7–9]. Therefore, it is important to understand hysteresis in models that describe neuronal activity, as this can help understand the processing mechanisms of different perception tasks.

Among the different perception task experiments that exhibit hysteresis we find the work established in [10]. In the aforementioned work, the authors developed several visual perception tasks to establish hysteresis as a function of an orientation disparity parameter. In the experimental settings, monocular tilted sinusoidal gratings were shown to the subjects. The orientation disparity between the gratings was increased by a fixed angle under different conditions. This setup led to subjects perceiving a single tilted grating due to stereoscopic tilt perception, or experiencing an alternation between the images perceived in the left and right eyes, known as binocular rivalry. The participants recorded the transition times when a single tilted grating changed

L. R. González-Ramírez (✉)
Escuela Superior de Física y Matemáticas, Instituto Politécnico Nacional, Unidad Profesional Adolfo López Mateos, Cd. de México, México
e-mail: lrgonzalezr@ipn.mx

J. J. Rosales García et al. (eds.), *Advances in Fractional Calculus*,
Modeling and Optimization in Science and Technologies 21,
https://doi.org/10.1007/978-3-031-83989-4_6

into two rivalrous gratings, and viceversa. The subjects had a resting time of 3-5 seconds between sequences. Noticeably, the transition between fusion and rivalrous gratings corresponded to a hysteresis loop between stereopsis and binocular rivalry as a function of orientation disparity. The same hysteresis effect was obtained even when the reaction time of the subjects was taken into account. Additionally, the authors established that the hysteresis loop was rate-dependent, that is, for a fast presentation rate of the tilted gratings a larger angle difference marked the transition between the two states. As the rate of the presentation decreased the angle difference that marked the transition between states also decreased. These results were consistent under different conditions of the experiment (100% and 25% of Michelson contrast). It was noted that larger hysteresis effects ocurred at 25% image contrast, while no hysteresis effect was observed at 100% and a slow presentation rate. Moreover, in the *blocked trial* experiments, the subjects were presented with the same type of image sequences without resting time in forward and reversed order. The aim was to explore the *subject's adaptation* to the experiment. The work established that in general, adaptation tends to decrease the hysteresis effect. The results obtained in the previous experiments are important for the work to be developed in this manuscript as they can be related to the effect of a fractional-order derivative operator. Indeed, we propose that the *subject's adaptation* may also be linked to a *memory* mechanism, as the subject is trained to experience the same experimental setting multiple times before the actual experiment. For full details of the previously mentioned experiments and results please see [10].

There have been efforts to mathematically model hysteresis at the neuronal level in an attempt to understand the biological mechanisms behind this phenomenon. At the neuronal activity level, the Wilson-Cowan model established the existence of hysteresis in the activity of homogeneous populations of excitatory and inhibitory neurons under convenient conditions and by modifying an external synaptic input [11, 12]. Hysteresis was hypothesized as a short-term neuronal memory mechanism motivated by the experimental observations found in [3]. The work developed in [11, 12] was later modified to establish the minimal conditions for modeling features in binocular rivalry [13]. Additionally, in [14] the authors developed a mathematical model to describe hysteresis in perception via a spiking neuron network. In their findings, they conclude that hysteresis is influenced by the rate of the external input and the weight of the excitatory synaptic interaction. Moreover in [15], a model was developed to describe binocular contrast, fusion, and rivalry, under different experimental conditions. The author established such model by considering two classes of inhibitory interneurons. It is important to note that in all of the previously mentioned works, the models were established using systems of ordinary (or spatially extended) differential equations of integer order.

Fractional-order models have been used to describe the non-local temporal nature of different systems and are linked to memory mechanisms of the underlying system [16, 24]. Therefore, it is important to include fractional-order derivatives in mathematical biology to explore the influence of memory on biological systems. However, it is crucial that the fractional-order model developed has a biological significance and is not just a mathematical extension of an integer-order biological system model. The goal of this work is to investigate hysteresis phenomena in a fractional-order neuronal model. In this way, we wish to establish a connection between the mathematical formalism and experimental observations. Therefore, we aim to analyze hysteresis in a fractional-order Wilson-Cowan model, previously motivated by biological factors to describe the effect of neuronal memory by incorporating a power-law kernel weighting past neuronal activity [18]. Recently, the study of fractional-order Wilson-Cowan formulations has revealed interesting emergent behaviors [19].

Fractional-order hysteresis loops have been previously studied mostly in the context of electrical circuits [20–22]. However, the authors are not aware of a qualitative or quantitative investigation of hysteresis in a fractional-order neuronal model. Therefore, this work aims to address this point. To that end, in Sect. 6.2, we introduce the model and motivate the existence of fractional-order hysteresis loops when considering a sigmoid activation function. These loops result from transitions between different stable states influenced by external synaptic input. This is done by linearizing the fractional-order system. Due to the limitations of this approach, we justify using a simplified Heaviside synaptic activation function. In Sect. 6.3, we explore the fractional-order model with a Heaviside activation function. In this simplified setting we obtain a fractional-order piecewise-linear system that under convenient conditions can be explicitly solved when considering a continuous periodic external synaptic input. In Sect. 6.4, different rates of the external input are considered to qualitatively compare results with experimental evidence. Additionally, in Sect. 6.5, we provide evidence that the fractional order can constrain neuronal hysteresis under convenient parameter choices. To complement our work, in Sect. 6.6, we obtain hysteresis loops when considering a sigmoid synaptic activation function. In the Discussion Section, we describe the impact of fractional order on neuronal phenomena, showing qualitatively consistent results with hysteresis effects in a visual perception experiment modeling the transition from fusion to binocular rivalry. Finally, in the Conclusion Section, we summarize our findings.

6.2 Hysteresis in the Fractional-Order Wilson-Cowan Model

We consider a fractional-order Wilson-Cowan model, which was biologically motivated in [18]. The model accounts for the memory-dependent neuronal activity of an excitatory and an inhibitory population. The model is established as:

$$\begin{aligned} {}_0D_t^{\alpha} u_e(t) &= -u_e(t) + F_e\left(w_{ee}u_e(t) - w_{ie}u_i(t) + P(t) - k_e\right) \\ {}_0D_t^{\alpha} u_i(t) &= -\delta u_i(t) + \delta F_i\left(w_{ei}u_e(t) - w_{ii}u_i(t) + Q(t) - k_i\right), \end{aligned} \tag{6.1}$$

where $u_j(t)$ accounts for the memory-dependent synaptic activity of the j-th population, $j = \{e, i\}$, where e is the excitatory and i is the inhibitory population, respectively. The function F_j determines the synaptic activation or response function which can be defined as a general sigmoid function. We propose $F_j(x) = \frac{1}{1+exp(-a_j x)}$, where the parameter $a_j > 0$ determines the steepness of the (synaptic) activation function, for $j = \{e, i\}$. The parameter $\delta > 0$ determines the change rate of the inhibitory population, and we consider $\delta \leq 1$. The parameters $w_{jk} > 0$, for $j, k = \{e, i\}$, determine the strength of the synaptic interaction from population j to population k. Additionally, the positive parameters k_j, for $j = \{e, i\}$, determine the synaptic threshold for the j-th population. The functions $P(t)$ and $Q(t)$ denote excitatory and inhibitory (synaptic) external inputs. In this formulation, we consider the Caputo fractional order derivative, with $0 < \alpha < 1$, and initial time at zero, as determined by the expression:

$$ {}_0D_t^\alpha f(t) := \frac{1}{\Gamma(1-\alpha)} \int_0^t \frac{f'(z)}{(t-z)^\alpha} dz, \tag{6.2} $$

for a real-valued function f such that both f and f' exist and are integrables in any open subinterval in $(0, \infty)$, and $\Gamma(z) := \int_0^\infty t^{z-1} e^{-t} dt$ is the Gamma function. Due to the biological motivation of model (6.1) that was developed in [18] the index α determines neuronal memory. Specifically, a low fractional order indicates greater neuronal memory, while a high fractional order indicates less memory. For properties of Caputo's fractional-order derivative, please refer to [23, 25].

We begin by examining the stability of the equilibrium solutions of the fractional-order system [26]. To do so, we assume there is no external inhibitory synaptic input, that is $Q(t) = 0$, and the excitatory external input is constant, i.e., $P(t) = P$, for $P \in \mathbb{R}$. In this way, the model determined by (6.1) can be considered as an autonomous fractional-order differential equation system (for a fixed value of P) and the equilibrium solutions are determined by $(u_e^*, u_i^*)^T$, where:

$$ \begin{aligned} u_e^* &= F_e(w_{ee}u_e^* - w_{ie}u_i^* + P - k_e) \\ u_i^* &= F_i(w_{ei}u_e^* - w_{ii}u_i^* - k_i). \end{aligned} \tag{6.3} $$

We note that the synaptic activation function satisfies $\frac{dF_j(u)}{du} = a_j F_j(u)(1 - F_j(u))$. Then, by linearizing at the equilibrium solution we obtain the Jacobian matrix:

$$ \begin{pmatrix} -1 + a_e w_{ee} u_e^*(1 - u_e^*) & -w_{ie} a_e u_e^*(1 - u_e^*) \\ \delta_i w_{ei} a_i u_i^*(1 - u_i^*) & -\delta_i - \delta_i w_{ii} a_i u_i^*(1 - u_i^*) \end{pmatrix}. \tag{6.4} $$

If both eigenvalues of the previous system satisfy $| Arg(\lambda) | > \frac{\alpha\pi}{2}$, then the equilibrium point is locally stable [25, 26]. By analyzing the nature of the nullclines it is possible to establish necessary and sufficient conditions for the existence of multiple equilibrium solutions, hysteresis, or oscillations. This has been done in an even more general system that includes a refractory period in [11]. By considering convenient parameters we provide evidence for the existence of fractional-order hysteresis loops

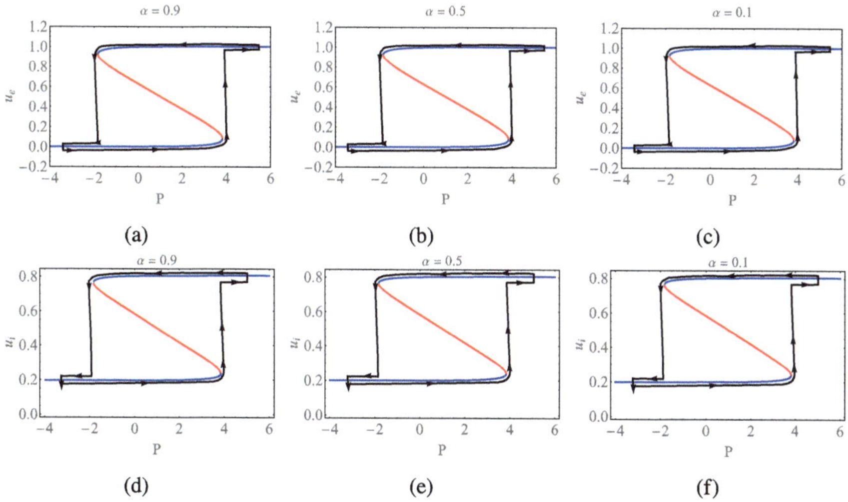

Fig. 6.1 Fractional-order hysteresis loops as the external synaptic input P is varied for different fractional orders. The blue branches indicate locally stable equilibrium points, while the red branch indicates locally unstable equilibrium points. Parameters used: $\delta = 1/3$, $w_{ee} = 20$, $w_{ei} = 4$, $w_{ie} = 12$, $w_{ii} = 2$, $k_e = 5$, $k_i = 1$ and $a_e = a_i = 1$. First row: excitatory population activity. Second row: inhibitory population activity. Fractional orders: (**a**)(**d**) $\alpha = 0.9$. (**b**)(**e**) $\alpha = 0.5$. (**c**)(**f**) $\alpha = 0.1$

of neuronal activity. Figure 6.1 shows evidence of hysteresis loops for the excitatory population (first row) and the inhibitory population (second row) at different fractional orders (first column: $\alpha = 0.9$, second column: $\alpha = 0.5$, and third column $\alpha = 0.1$). However, the qualitative understanding of the local stability of equilibrium solutions does not provide a clear effect of the influence of the fractional order on this phenomenon. Additionally, it is currently not possible to relate the effect of fractional order and external input rate to the experimental evidence of hysteresis in visual perception tasks.

In an effort to understand the effect of fractional order on hysteresis we consider the bifurcation diagram of neuronal activity when the parameter a_j of the synaptic activation function has increased. As depicted in Fig. 6.2, we still expect a hysteresis loop of neuronal activity when a_j is sufficiently large. However, there is not a clear effect of fractional order on the hysteresis. Nevertheless, we propose the use of a piecewise-linear activation function determined by a Heaviside function that can further simplify our study. Thus, we propose $F_j(x) = H(x)$, where $H(x) = 1$ for $x \geq 0$ and $H(x) = 0$ for $x < 0$. In this simplified setup, we can expect two equilibrium states for the excitatory population determined by one or zero and an unstable branch that provides a transition between these two states (see Fig. 6.2).

Under the previous considerations and by using a Heaviside activation function for both neuronal populations, the fractional-order system described by model (6.1) can be treated as a piecewise-linear system. Therefore, to take advantage of this

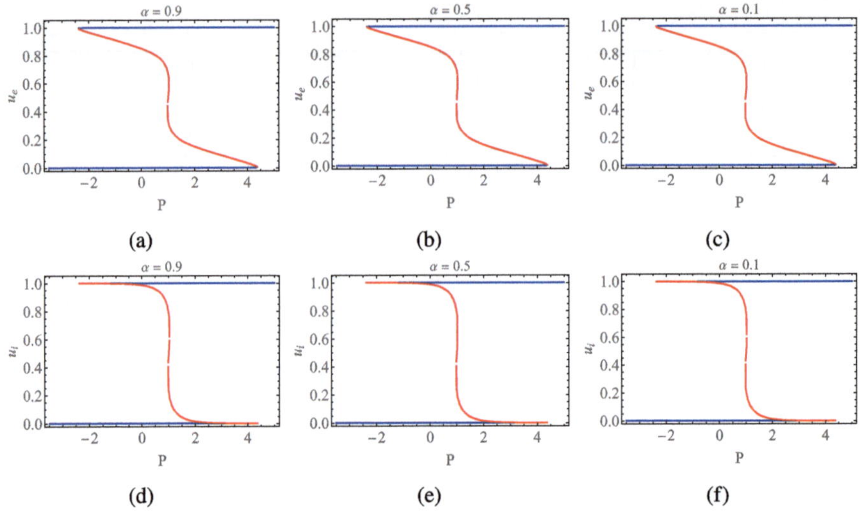

Fig. 6.2 Expected fractional-order hysteresis loops as the external synaptic input P is varied for different fractional orders and a very steep increase of the sigmoid activation function. The blue branches indicate locally stable equilibrium points, while the red branch indicates locally unstable equilibrium points. Parameters used: $\delta = 1/3$, $w_{ee} = 20$, $w_{ei} = 4$, $w_{ie} = 12$, $w_{ii} = 2$, $k_e = 5$, $k_i = 1$ and $a_e = a_i = 10$. First row: excitatory population activity. Second row: inhibitory population activity. Fractional orders: (**a**)(**d**) $\alpha = 0.9$. (**b**)(**e**) $\alpha = 0.5$. (**c**)(**f**) $\alpha = 0.1$

characteristic, we suggest treating the system as a non-homogeneous linear system determined by:

$$\begin{aligned} {}_0D_t^{\alpha} u_e(t) &= -u_e(t) + h_e(t) \\ {}_0D_t^{\alpha} u_i(t) &= -\delta u_i(t) + \delta h_i(t), \end{aligned} \tag{6.5}$$

where

$$h_j(t) = \begin{cases} 0, & \text{if } w_{ej}u_e(t) - w_{ij}u_i(t) + P_j(t) < k_j \\ 1, & \text{if } w_{ej}u_e(t) - w_{ij}u_i(t) + P_j(t) \geq k_j, \end{cases} \tag{6.6}$$

for $j = \{e, i\}$, $P_e(t) = P(t)$ and $P_i(t) = Q(t)$. This is a linear system, and once $h_j(t)$ is understood, we can find an explicit solution for this fractional-order system [24, 25]. Additionally, with sufficiently low external excitatory and inhibitory inputs, we can expect that both populations are below the synaptic threshold, and $h_e(t) = h_i(t) = 0$ for $t \in [0, t^\star]$, for a sufficiently small $t^\star > 0$. In this case, we obtain the explicit solutions:

$$\begin{aligned} u_e(t) &= u_0^e E_\alpha(-t^\alpha) \\ u_i(t) &= u_0^i E_\alpha(-\delta t^\alpha), \end{aligned} \tag{6.7}$$

where u_0^e and u_0^i are the excitatory and inhibitory initial conditions and $E_\alpha(t)$ is a one-parameter Mittag-Leffler function. For properties of Mittag-Leffler functions please see [23, 25]. As time increases, and potentially the external input increases, we may expect that both neuronal populations will exceed the synaptic threshold. Therefore, in general, the solution to system (6.5) is determined by:

$$\begin{aligned} u_e(t) &= u_0^e E_\alpha(-t^\alpha) + \int_0^t h_e(t-s)s^{\alpha-1}E_{\alpha,\alpha}(-s^\alpha)ds \\ u_i(t) &= u_0^i E_\alpha(-\delta t^\alpha) + \int_0^t \delta h_i(t-s)s^{\alpha-1}E_{\alpha,\alpha}(-\delta s^\alpha)ds, \end{aligned} \tag{6.8}$$

where $E_{\alpha,\alpha}(t)$ is a two-parameter Mittag-Leffler function [23, 25].

6.3 Hysteresis Loops for a Heaviside Activation Function

Considering initial conditions $u_0^e = u_0^i = 0$ and the previous assumptions, we develop a scheme to create explicit fractional-order solutions. We assume that the inhibitory external input is set to zero in order to achieve a scenario qualitatively similar to the one shown in Figs. 6.1 and 6.2. For the remainder of the document, when we mention an external input, we are referring to an excitatory external synaptic input. We propose to vary the increase/decrease of the excitatory external input using a continuous piecewise-linear function. Considering a convenient initial increase of the external input, we expect the synaptic excitatory input to exceed the synaptic threshold, leading to activation of the excitatory population. This will produce a similar process for the inhibitory population. A subsequent decrease in the excitatory external input will result in the excitatory synaptic input falling below the synaptic threshold, followed by a decrease in the excitatory population activity. Similar effects can be expected in the inhibitory population. In order to create this loop, the external input must produce an increase in the excitatory/inhibitory synaptic input sufficient to surpass the excitatory/inhibitory threshold, as well as a subsequent decrease in the excitatory/inhibitory synaptic input below the excitatory/inhibitory threshold. In this process, the excitatory dynamics precede to the inhibitory dynamics. The activation/deactivation of the synaptic excitatory/inhibitory input determines the times at which the functions $h_j(t)$ activate and deactivate, and therefore, we can obtain the activation/deactivation of the non-homogeneous term of the fractional-order system. With this observation, and assuming that the activation and deactivation times of each population do not occur simultaneously, we can obtain the explicit solutions:

$$u_e(t) = \begin{cases} 0, & \text{for } 0 \le t \le t_1 \\ (t-t_1)^\alpha E_{\alpha,\alpha}(-(t-t_1)^\alpha), & \text{for } t_1 \le t \le t_3, \\ (t-t_1)^\alpha E_{\alpha,\alpha}(-(t-t_1)^\alpha) - (t-t_3)^\alpha E_{\alpha,\alpha}(-(t-t_3)^\alpha), & \text{for } t_3 \le t \le t_5, \\ (t-t_5)^\alpha E_{\alpha,\alpha}(-(t-t_5)^\alpha) + & \\ (t-t_1)^\alpha E_{\alpha,\alpha}(-(t-t_1)^\alpha) - (t-t_3)^\alpha E_{\alpha,\alpha}(-(t-t_3)^\alpha), & \text{for } t_5 \le t \le t_7, \\ (t-t_5)^\alpha E_{\alpha,\alpha}(-(t-t_5)^\alpha) - (t-t_7)^\alpha E_{\alpha,\alpha}(-(t-t_7)^\alpha) + & \\ (t-t_1)^\alpha E_{\alpha,\alpha}(-(t-t_1)^\alpha) - (t-t_3)^\alpha E_{\alpha,\alpha}(-(t-t_3)^\alpha), & \text{for } t_7 \le t \le t_9, \\ \dots \end{cases} \tag{6.9}$$

$$u_i(t) = \begin{cases} 0, & \text{for } 0 \le t \le t_2 \\ \delta(t-t_2)^\alpha E_{\alpha,\alpha}(-\delta(t-t_2)^\alpha), & \text{for } t_2 \le t \le t_4, \\ \delta(t-t_2)^\alpha E_{\alpha,\alpha}(-\delta(t-t_2)^\alpha) - \delta(t-t_4)^\alpha E_{\alpha,\alpha}(-\delta(t-t_4)^\alpha), & \text{for } t_4 \le t \le t_6, \\ \delta(t-t_6)^\alpha E_{\alpha,\alpha}(-\delta(t-t_6)^\alpha) + & \\ \delta(t-t_2)^\alpha E_{\alpha,\alpha}(-\delta(t-t_2)^\alpha) - \delta(t-t_4)^\alpha E_{\alpha,\alpha}(-\delta(t-t_4)^\alpha), & \text{for } t_6 \le t \le t_8, \\ \delta(t-t_6)^\alpha E_{\alpha,\alpha}(-\delta(t-t_6)^\alpha) - \delta(t-t_8)^\alpha E_{\alpha,\alpha}(-\delta(t-t_8)^\alpha) + & \\ \delta(t-t_2)^\alpha E_{\alpha,\alpha}(-\delta(t-t_2)^\alpha) - \delta(t-t_4)^\alpha E_{\alpha,\alpha}(-\delta(t-t_4)^\alpha), & \text{for } t_8 \le t \le t_{10}, \\ \dots \end{cases} \tag{6.10}$$

where the t_j's determine the activation/deactivation times of the synaptic external input and are determined by the threshold equations:

$$w_{ee}u_e(t_{2n+1}) - w_{ie}u_i(t_{2n+1}) + P(t_{2n+1}) = k_e, \tag{6.11}$$

for $n \in \mathbb{N} \cup \{0\}$, and

$$w_{ei}u_e(t_{2n}) - w_{ii}u_i(t_{2n}) = k_i, \tag{6.12}$$

for $n \in \mathbb{N}$.

To numerically validate the previous solutions we propose a piecewise-linear periodic external input determined by:

$$P(t) = \begin{cases} P_1(t), & \text{for } 0 \le t \le p \\ P_2(t), & \text{for } p \le t \le 2p, \\ P_1(t-2p), & \text{for } 2p \le t \le 3p \\ P_2(t-2p), & \text{for } 3p \le t \le 4p \\ P_1(t-4p), & \text{for } 4p \le t \le 5p \\ P_2(t-4p), & \text{for } 5p \le t \le 6p, \\ \dots \end{cases} \tag{6.13}$$

where $P_1(t) = mt - 2\Delta$, $P_2(t) = -m(t-p) + k_e + 2\Delta$. We assume that $\Delta \ge 0$ is a sufficiently large parameter that permits the activation/deactivation of both populations. Also, $p = \frac{k_e+4\Delta}{m} > 0$ determines one half of the period of the external input.

The parameter $m > 0$ determines the rate of increase of the external input (correspondingly $-m$ determines the rate of decrease). Thus, a change of rate $m > 0$ determines different periods of the external input. We choose this form of the external input to qualitatively compare the effect of the rate of change of the external input and the effect of the fractional-order derivative on hysteresis as determined in [10]. In particular, and to simplify our analysis we consider three scenarios for the parameter m. We consider that $m = 1$ is a medium external input rate, $m = 0.2$ is a slow external input rate and $m = 2$ is a fast external input rate. We then proceed to numerically solve the threshold conditions determined by Eqs. (6.11) and (6.12) by considering (6.13). This is done developing a numerical routine in the software Mathematica [27].

We first obtain the explicit solutions motivated by the bifurcation diagrams obtained in Fig. 6.2. To do so, we numerically obtain the time points t_k at which synaptic activation or deactivation occurs, satisfying Eqs. (6.11) or (6.12). To do so, we consider a sufficiently large range of the external synaptic input, determined by a value of $\Delta = 4$. With this information, we can plot the explicit solutions determined by Eq. (6.9) and Eq. (6.10). In Fig. 6.3, we analyze the impact of fractional order on neuronal activities for different rates of the excitatory external input. The columns in Fig. 6.3 represent different fractional orders (first column: $\alpha = 0.99$, second column: $\alpha = 0.6$, and third column: $\alpha = 0.01$). The rows in Fig. 6.3 represent different external input rates (first row: slow input rate, second row: medium input rate, and third row: fast input row). Our findings show that both the fractional order and the external input rate have a qualitative effect on neuronal population activities. Specifically, the fractional order diminishes the increase of activity of both neuronal populations. As the fractional order decreases (and the memory index increases), the increase of the neuronal population due to the periodic external input is diminished. Also, there is a non-linear effect of the external input rate on the qualitative properties of the neuronal population. Slow input rates tend to have a more abrupt transition toward the "excited" state (i.e., high neuronal activity) or the "rest" state (i.e., low neuronal activity) compared to faster external input rates, which tend to have a slower transition toward an increased or decreased neuronal activity state. However, the effect of the external input rate is diminished as the fractional order is decreased. For sufficiently low fractional order, the neuronal activity shows no apparent distinction between different external input rates. Overall, we observe qualitative differences in the effect of fractional order on neuronal activity. Additionally, we expect an external input rate and fractional order dependence on neuronal hysteresis, which we will further analyze in the following sections.

We now take advantage of our explicit solutions to obtain hysteresis loops of neuronal activities. In Fig. 6.4, we plot the hysteresis loop of the excitatory population as a function of the chosen periodic external input. The analysis is conducted for different external input rates (first column: slow external input rate, second column: medium external input rate, and third column: fast external input rate) and for both neuronal populations (first row: excitatory activity, and second row: inhibitory activity). Additionally, we explore the impact of different fractional orders on the hysteresis loop : $\alpha = 0.99$ (black curves), $\alpha = 0.6$ (cyan curves), and $\alpha = 0.01$ (magenta

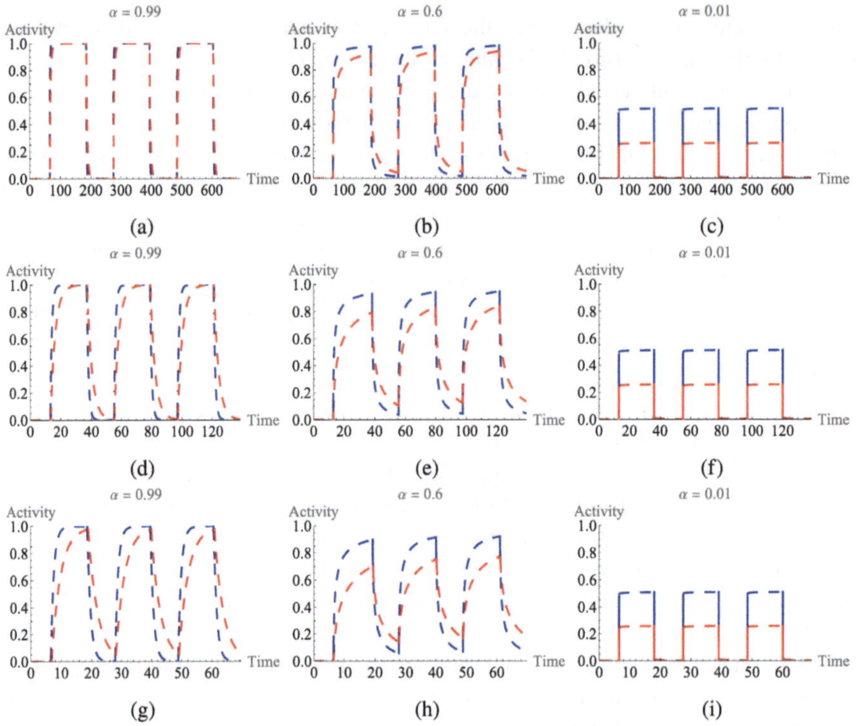

Fig. 6.3 Excitatory and inhibitory neuronal activities as a function of a periodic external synaptic input. The blue dashed curves indicate excitatory activity, and the red dashed curves indicate inhibitory activity. Parameters used: $\delta = 1/3$, $w_{ee} = 20$, $w_{ei} = 4$, $w_{ie} = 12$, $w_{ii} = 2$, $k_e = 5$, and $k_i = 1$. First row: slow external input rate ($m = 0.2$). Second row: medium external input rate ($m = 1$). Third column: fast external input rate ($m = 2$). Fractional orders: **(a)**(**d**)(**g**) $\alpha = 0.99$. (**b**)(**e**)(**f**) $\alpha = 0.6$. (**c**)(**f**)(**i**) $\alpha = 0.01$

curves). Moreover, the explicit fractional order solutions allow us to determine the exact times of activation and deactivation of synaptic inputs. By analyzing these times we observe that they determine the increase/decrease of the loop (see the colored dots in the hysteresis loops). We observe a non-linear effect of fractional order on neuronal hysteresis. As the fractional order decreases ($\alpha = 0.6$), the excitatory and inhibitory external input differences slightly increase (differences in the external input at the activation and deactivation times for the excitatory and inhibitory populations). However, for sufficiently low fractional order ($\alpha = 0.01$), the neuronal input difference decreases. Additionally, the external input rate appears to have a slight qualitative effect on the shape of the hysteresis loops. Generally, as the fractional order decreases, the upper branch of the loop tends to diminish. This is not necessarily due to a decrease in the stable equilibrium in the upper branch, but rather an effect of the slower-power law tendency caused by the fractional-order derivative.

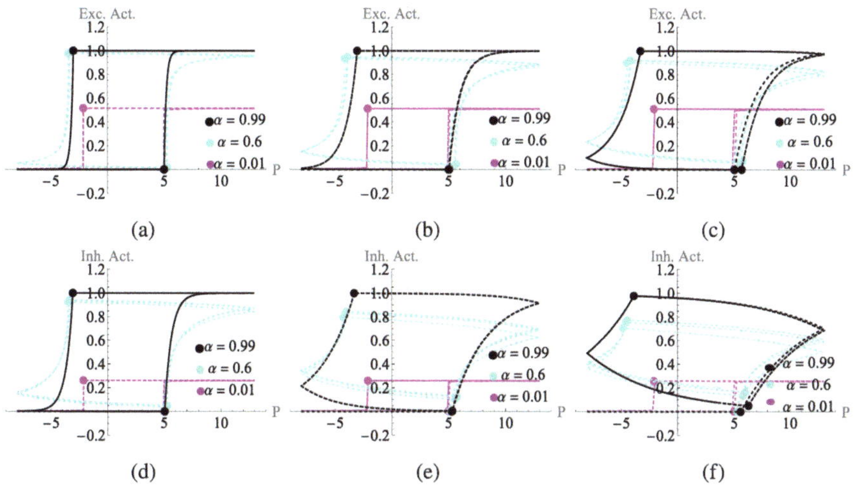

Fig. 6.4 Fractional-order hysteresis loops as the external synaptic input P is varied for different fractional orders. The initial effect of fractional order is to increase the hysteresis lag. For sufficiently low fractional order the hysteresis lag is decreased. Parameters used: $\delta = 1/3$, $w_{ee} = 20$, $w_{ei} = 4$, $w_{ie} = 12$, $w_{ii} = 2$, $k_e = 5$, and $k_i = 1$. First row: excitatory population activity. Second row: inhibitory population activity. Fractional orders: $\alpha = 0.99$ (black dashed curves), $\alpha = 0.6$ (cyan dashed curves) and $\alpha = 0.01$ (magenta dashed curves). First column: slow input rate. Second column: medium input rate. Third column: fast input rate

This effect results in a smaller excitation on the neuronal activity. The lower branch of the loop is generally not severely affected by fractional order.

6.4 Adaptation Effects in Visual Perception Tasks are Qualitatively Consistent with Effects Due to Sufficiently Low Fractional-Order

The information observed from Fig. 6.4 may be further quantified in order to relate to the experimental findings observed in a visual perception task in which an adaptation of the task produced a decrease in the hysteresis lag. However, this effect cannot be directly determined from Fig. 6.4. Therefore, numerical routines were developed to obtain the neuronal external input difference (i.e., the difference in the external input at the activation/deactivation times) that establishes the hysteresis lag. We analyzed the mean external input difference as a function of fractional order and external input rate. The results of this analysis are shown in Fig. 6.5, where we plotted the external input difference as a function of the reciprocal of the input rate. The reason for choosing this presentation is to directly relate to the experimental findings shown in Figs. 6.4 and 6.5 of the work developed in [10], in which a slower input rate produced a decrease in the external input difference. In our findings, we observed that

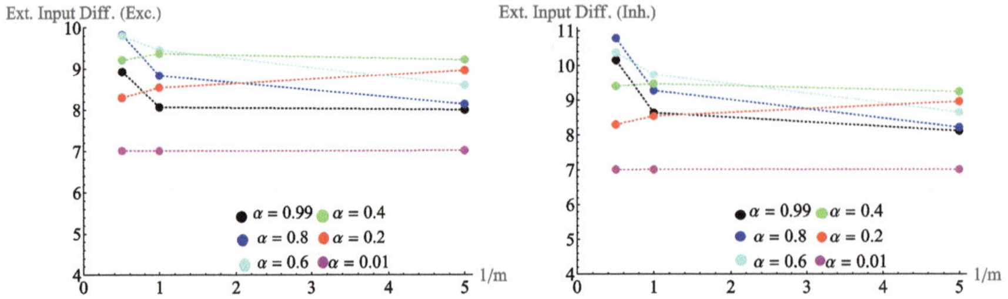

(a) (b)

Fig. 6.5 Excitatory and inhibitory external input difference (hysteresis lag) as a function of the reciprocal of the external input rate ($1/m$). The external input rates are determined by: $m = 0.2$ is the slow input rate ($1/m = 5$), $m = 1$ is the medium input rate ($1/m = 1$), and $m = 2$ is the fast input rate ($1/m = 0.5$). Parameters used: $\delta = 1/3$, $w_{ee} = 20$, $w_{ei} = 4$, $w_{ie} = 12$, $w_{ii} = 2$, $k_e = 5$ and $k_i = 1$. **(a)** Excitatory external input difference. **(b)** Inhibitory external input difference. Fractional orders: $\alpha = 0.99$ (black curve), $\alpha = 0.8$ (blue curves), $\alpha = 0.6$ (cyan curves), $\alpha = 0.4$ (green curves), $\alpha = 0.2$ (red curves) and $\alpha = 0.01$ (magenta curves)

high fractional orders have the same rate dependency as the experimental evidence. However, the initial effect of fractional order is to *increase* the hysteresis lag. For sufficiently low fractional order ($\alpha = 0.01$), we observed a decrease in the hysteresis lag. This scenario produces consistency with the reported experimental findings in the *blocked trial* experimental setting. This effect occurs for both neuronal population activities. Therefore, the adaptation of the visual task is qualitatively consistent with our findings only for sufficiently low fractional order (and high memory index).

In the following analysis, we will examine the impact of fractional order on neuronal populations under a different scenario. In Fig. 6.6(a),(b), we show a potential scenario of neuronal hysteresis where five steady states coexist, two of which are stable and three are unstable. There is no noticeable difference in the effect of fractional order on hysteresis loops. Consequently, we proceed to investigate the effect of the fractional order on neuronal differences under various external input rates. The results we obtained are qualitatively similar to those in Fig. 6.5. There is a non-linear effect of fractional order on excitatory and inhibitory external input differences. Initially, fractional order increases the external input difference. There is a consistent decrease in fractional order depending on the input rate. A slower input rate results in a decrease in the input difference. For a sufficiently low fractional order ($\alpha = 0.2$), the effect of fractional order leads to a decrease of the input difference. However, in this instance, the rate does not notably influence the dynamics. In both scenarios shown in Figs. 6.5 and 6.6, we consistently observe a decrease in hysteresis lag as a function of input rate, but only for sufficiently high fractional order (i.e., $\alpha = 0.99$ and $\alpha = 0.8$). Furthermore, the diminishing effect due to adaptation is qualitatively similar for sufficiently low fractional orders (high memory index). In other words, we can qualitatively replicate the experimental findings of adaptation on a visual task but **only for sufficiently low fractional orders** and high memory index. We also remark that the numerical results shown in the previous sections were verified

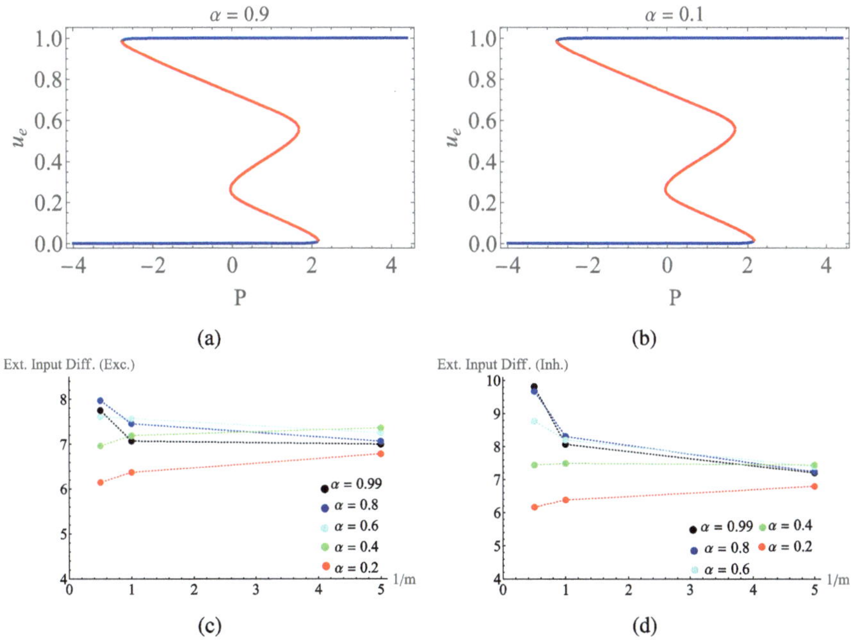

Fig. 6.6 **a**, **b** Expected fractional-order excitatory hysteresis loops as the external synaptic input P is varied for different fractional-orders and a very steep increase of the sigmoid activation function. The blue branches indicate locally stable equilibrium points. The red branch indicates locally unstable equilibrium points. For some values of P, there is coexistence of 5 equilibrium points. Fractional orders: (**a**) $\alpha = 0.9$ and (**b**) $\alpha = 0.1$. (**c**),(**d**) Excitatory and inhibitory external input difference (hysteresis lag) as a function of the reciprocal of the external input rate ($1/m$) when considering a Heaviside activation function. The external input rates are determined by: $m = 0.2$ is the slow input rate ($1/m = 5$), $m = 1$ is the medium input rate ($1/m = 1$) and $m = 2$ is the fast input rate ($1/m = 0.5$). (**c**) Excitatory external input difference. (**d**) Inhibitory external input difference. Fractional orders: $\alpha = 0.99$ (black curve), $\alpha = 0.8$ (blue curve), $\alpha = 0.6$ (cyan curve), $\alpha = 0.4$ (green curve), and $\alpha = 0.2$ (red curve). Parameters used: $\delta = 1/3$, $w_{ee} = 13.5$, $w_{ei} = 9$, $w_{ie} = 6.5$, $w_{ii} = 2$, $k_e = 3.2$, and $k_i = 2.7$

by numerically solving model (6.1) developing a routine in [28] using a fractional Adams-Bashforth-Moulton method [17].

6.5 Fractional Order can Constrain Neuronal Hysteresis

We now proceed to analyze a different example compared to the previous cases. Here, we consider a scenario where the inhibitory threshold activation and deactivation happen at the same time. Consequently, we cannot directly use our explicit solutions to obtain the hysteresis loops. In Fig. 6.7(a),(b), we show the bifurcation diagram as a function of the external input for a fractional order of $\alpha = 0.9$ for both excitatory and inhibitory populations. We observe that the lower inhibitory stable branch collapses with an unstable branch. When we attempt to utilize the explicit solution established in Eqs. (6.9) and (6.10), we need to first analyze the threshold conditions established in Eqs. (6.11) and (6.12). We numerically obtained the activation of the excitatory activity (marked as a blue circle labeled as t_1 in Fig. 6.7(c)–(e). As a consequence, this produces an activation of the inhibitory population (marked as a red circle labeled as t_2 in Fig. 6.7(c)–(e). However, upon analyzing the inhibitory synaptic input in Fig. 6.7(e), we observe that t_2 is also the point in which inactivation of the inhibitory synaptic input occurs, and the inhibitory population cannot be above the threshold for a non-trivial time interval. Therefore, our explicit solution established in Eq. (6.10) cannot accurately describe this scenario. Thus, we proceed to analyze numerical simulations of this case. Figure 6.8 displays our results, where we fix a medium external input rate (i.e., $m = 1$) and vary the fractional order (first row: $\alpha = 0.99$, second row: $\alpha = 0.6$, and third row: $\alpha = 0.2$). The neuronal activities are shown in the first column, the excitatory hysteresis loop as a function of the external input is shown in the second column and the inhibitory loop is shown in the third column. We observe that for high fractional order (e.g., $\alpha = 0.99$ and $\alpha = 0.6$), we obtain periodic neuronal activity and hysteresis loops, similar to the cases explored in previous sections. However, there is a significant effect of fractional order on the inhibitory neuronal population activity as it is considerably diminished. The excitatory neuronal loop is not noticeably affected by the external input, whereas the inhibitory loop is diminished as the fractional order is decreased. Moreover, for sufficiently low fractional order (e.g., $\alpha = 0.2$), the inhibitory activity cannot increase and it remains in the rest state regardless of the influence of the external input. In this case, there is a slight diminishing effect on the excitatory activity due to the fractional order, whereas the inhibitory activity is severely affected.

6.6 Hysteresis Loops for a Sigmoid Activation Function

In this section, to complement our work, we conducted a numerical analysis of solutions for model (6.1), considering a general sigmoid activation function. In Fig. 6.9, we observed neuronal hysteresis using the same parameters as those used in Fig. 6.4. However, since we do not have explicit solutions, we cannot directly determine the time points at which activation or deactivation of activity occurs. Therefore, we are unable to quantitatively measure the hysteresis lag. However, a qualitative analysis revealed both similarities and differences with the case of a Heaviside activation function. For sufficiently high fractional orders (e.g., $\alpha = 0.99$ and $\alpha = 0.6$), there is no clear distinction in the effect of fractional order on hysteresis loops. However, as the fractional order decreases, the upper branch of the loop diminishes. In comparison to the Heaviside case, there is no increase in hysteresis lag. Additionally, there are also similarities between the two possible scenarios of activation functions. For example, with sufficiently low fractional order (e.g., $\alpha = 0.1$), the external

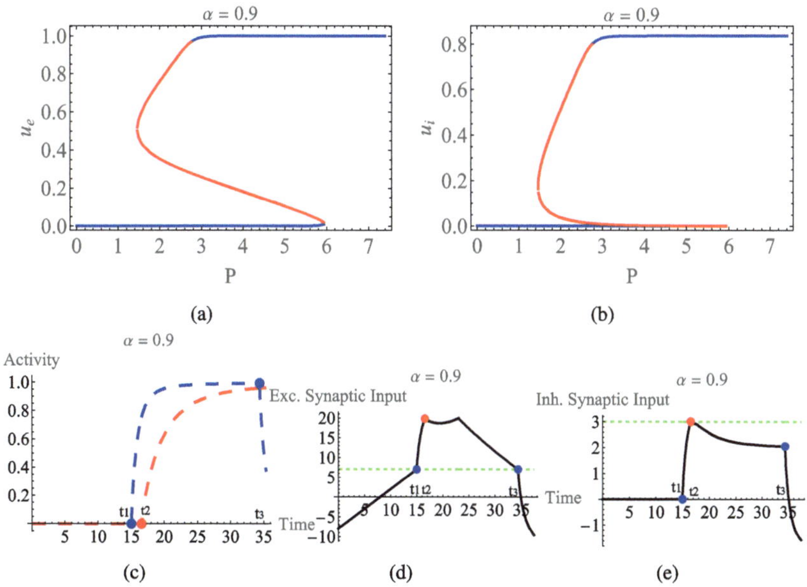

Fig. 6.7 Expected fractional-order hysteresis loops as the external synaptic input P is varied for different fractional-orders and a steep increase of the sigmoid activation function. The blue branches indicate locally stable equilibrium point and the red branch indicates locally unstable equilibrium points. Fractional orders: (**a**) $\alpha = 0.9$ and (**b**) $\alpha = 0.1$. Considering a fractional order of $\alpha = 0.9$, we analyze the explicit solution (**c**), the excitatory synaptic input (**d**) and the inhibitory synaptic input (**e**). Following the solution construction scheme, the excitatory population (blue dashed curve) activates at t_1. This can be observed when excitatory synaptic input reaches the synaptic threshold (black curve in (**d**) that reaches the green dashed curve at the point marked with a blue dot and labeled as t_1). This in turn produces the activation of inhibition (red dashed curve) at t_2. However, inhibitory synaptic input (black curve in (**d**) that reaches the green dashed curve at t_2 marked with a red point) does not become suprathreshold for a non-trivial time interval. The blue points indicate activation/deactivation points of the excitatory activity whereas red points indicate activation/deactivation points of the inhibitory activity. Parameters used: $\delta = 1/3$, $w_{ee} = 15$, $w_{ei} = 4$, $w_{ie} = 12$, $w_{ii} = 2$, $k_e = 7$, and $k_i = 3$

input rate does not affect the hysteresis loop. Therefore, as a general observation in the case of a more general activation function, we may also expect a non-linear, rate-dependent, and fractional-order dependent effect on neuronal hysteresis (see Fig. 6.9).

6.7 Discussion

Fractional-order derivatives have been associated to memory events in various phenomena (e.g., [16, 24]), and have been linked to neuronal memory processes. Neuronal memory refers to the fractional-order operator taking into account previous neuronal activity [18]. In our study, we aimed to connect our analytical findings of

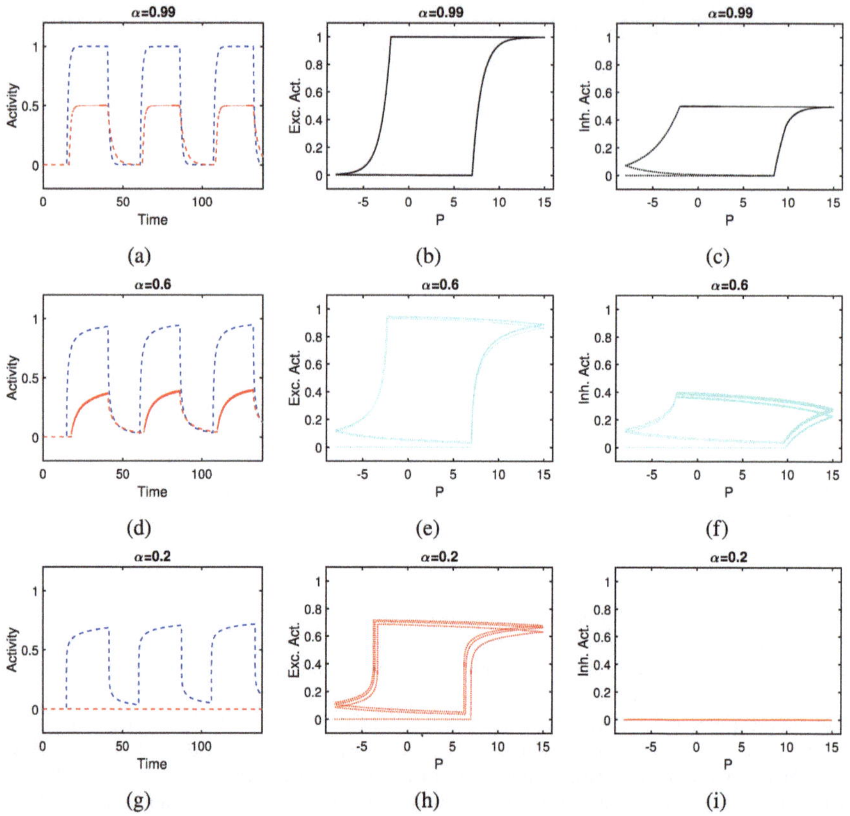

Fig. 6.8 Numerical simulations of neuronal activity as a function of external synaptic input. First column: excitatory population activities (blue dashed curves) and inhibitory population activities (red dashed curves). Second column: excitatory activity as a function of external input. Third column: inhibitory activity as a function of external input. Fractional orders: **(a)**–**(c)** $\alpha = 0.99$. **(d)**–**(f)** $\alpha = 0.6$. **(g)**–**(i)** $\alpha = 0.2$. Parameters used: $\delta = 1/3$, $w_{ee} = 15$, $w_{ei} = 4$, $w_{ie} = 12$, $w_{ii} = 2$, $k_e = 7$, and $k_i = 3$

explicit neuronal activity to experimental observations of hysteresis during visual perception tasks, particularly related to the hysteresis between the stereoscopic perception of tilt and binocular rivalry. We choose this experimental evidence due to the completeness of their setting. In this experiment, the rate dependency of the external input determined a decrease or increase in hysteresis lag. A slow presentation rate was associated with a lower hysteresis lag. However, the results were also affected by the image contrast. For 25% image contrast, the previous results were consistent, but for 100% contrast, there was a nonlinear relation between input rate and hysteresis. The *blocked trial* setting of the experiment may be viewed as a memory phenomenon, as participants were repeatedly shown the experimental images before

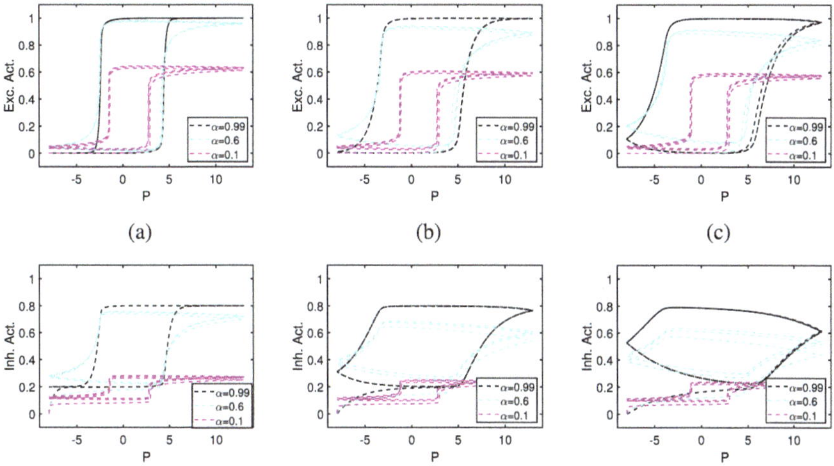

(a) (b) (c) (d) (e) (f)

Fig. 6.9 Fractional-order hysteresis loops as the external synaptic input P is varied for different fractional-orders considering a sigmoid activation function. The effect of fractional order is to decrease the hysteresis lag. Parameters used: $\delta = 1/3$, $w_{ee} = 20$, $w_{ei} = 4$, $w_{ie} = 12$, $w_{ii} = 2$, $k_e = 5$, and $k_i = 1$. First row: excitatory population activity. Second row: inhibitory population activity. Fractional orders: $\alpha = 0.99$ (black dashed curves), $\alpha = 0.6$ (cyan dashed curves) and $\alpha = 0.1$ (magenta dashed curves)

the actual experiment as a preparation procedure. The results of the experiment indicate that the *blocked trial* setting reduced the hysteresis lag (see Fig. 6 in [10]). Therefore, a plausible memory phenomenon in a visual perception task is associated with a decrease in hysteresis lag. Our modeling results agree with some of these observations and disagree with others. For example, the mathematical model implies that for sufficiently high fractional orders we find a qualitative consistency with the results shown in the 25% image contrast: as the external input rate is decreased the hysteresis lag is decreased. However, as the fractional order is decreased, the rate effect on hysteresis is not noticeable. Interestingly, the results found for low fractional orders are more consistent with the results for 100% image contrast (see Fig. 3 in [10]). Additionally, our results show that for specific parameters, it is possible to avoid the hysteresis loop for the inhibitory population, which occurs with the experimental setting at the 100% image contrast and slow input rate. This suggests that fractional order may potentially incorporate specific aspects of neuronal information. The *blocked trial* setting and its experimental results agree overall with the effect of low fractional order on neuronal hysteresis (see Fig. 6 in [10]). It is unclear why this effect only occurs for sufficiently low fractional orders, as higher fractional orders seem to cause a counterintuitive increase in hysteresis lag. This increase in hysteresis lag might be justified by the overall decrease in neuronal activity, which potentially adjust the overall memory effect. Interestingly, simulations using the sigmoid activation function do not show this initial increase in hysteresis lag, suggesting it could also be an artifact of the simplified activation function. However, our findings suggest that qualitatively similar results to the experimental *adaptation* are achieved for sufficiently low fractional order, corresponding to a higher memory index. Nonetheless, given the specific details previously described, further research using more specific fractional-order neuronal models is necessary to analyze the exact impact of neuronal memory on hysteresis phenomena.

6.8 Conclusions

In this work, we established explicit fractional-order solutions in a fractional-order Wilson-Cowan model of neuronal activity, taking into account a Heaviside synaptic activation function with a periodic external input and in the absence of inhibitory external input. We observed intriguing non-linear effects resulting from the fractional order on neuronal hysteresis. Specifically, we found evidence of a non-linear, rate-dependent, and fractional order-dependent impact on neuronal activity. For sufficiently large fractional orders (near one) we noted a decrease in hysteresis lag as the external input rate was decreased. Fast external inputs led to higher hysteresis lag, while slower rates resulted in lower hysteresis lag. However, an overall increase in hysteresis lag was observed as the fractional order decreased. On the other hand, for sufficiently low fractional orders (that depends on the parameter choice) an overall decrease in hysteresis lag was observed. Additionally, when the fractional order is low, the effect of the rate of external input is not substantial. Even though for low fractional order there is no considerable effect due to the rate of external input, we can recover an overall decrease of the hysteresis loop. We also found that for more complex parameter configurations, there can be simultaneous activation and deactivation of the inhibitory synaptic threshold. In such cases, the effect of the fractional order is more pronounced and can, for sufficiently low fractional orders, reduce or eliminate hysteresis in the inhibitory population. In summary, our findings indicate qualitatively consistent results of the memory effect (*subject's adaptation*) in agreement with high neuronal memory index in visual perception tasks and shed light on a complex relation between fractional order and neuronal activity.

Acknowledgements We thank the anonymous referees for their helpful comments and feedback.
Competing Interests This study was funded by SIP-IPN 20241172.

References

1. Noori HR (2014) Examples of Hysteresis Phenomena in Biology. In: Hysteresis phenomena in biology, pp 34–35. Springer, Heidelberg
2. Mayergoyz ID (2003) Mathematical models of hysteresis and their applications. Academic Press
3. Fender D, Julesz B (1967) Extension of panum's fusional area in binocularly stabilized vision. J Opt Soc America 57(6):819–830. https://doi.org/10.1364/JOSA.57.000819
4. Williams D, Phillips G, Skelur R (1986) Hysteresis in the perception of motion direction as evidence for neural cooperativity. Nature 324:253–255. https://doi.org/10.1038/324253a0
5. Poltoratski S, Tong F (2014) Hysteresis in the dynamic perception of scenes and objects. J Exp Psychol Gen. https://doi.org/10.1037/a0037365
6. Liaci E, Fischer A, Atmanspacher H, Heinrichs M, Terbatz van Elst L, Poltoratski S, Tong F (2018) Positive and negative hysteresis effects for the perception of geometric and emotional ambiguities. PLOS ONE 13(9):e0202398. https://doi.org/10.1037/a0037365, https://doi.org/10.1371/journal.pone.0202398
7. Hubel DH, Wiesel TN (1970) Stereoscopic vision in macaque monkey. Cells sensitive to binocular depth in area 18 of the macaque monkey cortex. Nature 225(5227):41–2. https://doi.org/10.1038/225041a0
8. Sayal A, Sousa T, Duarte JV, Costa GN et al (2020) Identification of competing neural mechanisms underlying positive and negative perceptual hysteresis in the human visual system. NeuroImage 221:117153. https://doi.org/10.1016/j.neuroimage.2020.117153
9. Kleinschmidt A, Büchel C, Hutton C, Friston KJ, Frackowiak SJ (2002) The neural structures expressing perceptual hysteresis in visual letter recognition. Neuron 34(4). https://doi.org/10.1016/S0896-6273(02)00694-3
10. Buckthought A, Kim J, Wilson HR (2008) Hysteresis effects in stereopsis and binocular rivalry. Vis Res 48(6). https://doi.org/10.1016/j.visres.2007.12.013D
11. Wilson HR, Cowan JD (1972) Excitatory and inhibitory interactions in localized populations of model neurons. Biophysics J 12(1):1–24. https://doi.org/10.1016/S0006-3495(72)86068-5
12. Wilson HR, Cowan JD (1973) A mathematical theory of the functional dynamics of cortical and thalamic nervous tissue. Kybernetik. https://doi.org/10.1007/BF00288786
13. Wilson HR (2007) Minimal physiological conditions for binocular rivalry and rivalry memory. Vis Res 47:2741–2750
14. You H, Meng Y, Huan D, Wang DH (2011) The neural dynamics for hysteresis in visual perception. Neurocomputing 74:3502–3508
15. Wilson HR (2017) Binocular contrast, stereopsis, and rivalry: toward a dynamical synthesis. Vis Res 140:89–95
16. Du M, Wang Z, Hu H (2013) Measuring memory with the order of fractional derivative. Sci Rep 3:3431. https://doi.org/10.1038/srep03431
17. Diethelm K, Ford NJ, Freed AD (2004) Detailed error analysis for a fractional Adams method. Numer Algorithms 36
18. González-Ramírez LR (2024) A fractional-order Wilson-Cowan formulation of cortical disinhibition. J Comput Neurosci
19. Mondal A, Kaslik E, Sharma SK, Chakraborty C, Aziz-Alaoui MA (2024) Emergent dynamics in fractional-order Wilson-Cowan neural network systems. Chaos Solitons Fractals 181:114687
20. Caputo M, Fabrizio M (2017) On the notion of fractional derivative and applications to the hysteresis phenomena. Meccanica 52:3043–3052. https://doi.org/10.1007/s11012-017-0652-y
21. Hernández-Balaguera E, Arredondo B, del Pozo G, Romero B (2020) Exploring the impact of fractional-order capacitive behavior on the hysteresis effects of perovskite solar cells: a theoretical perspective. Commun Nonlinear Sci Numer Simul 90:105371
22. Liu L, Yun H, Li Q, Ma X, Chen SL Shen J (2020) Fractional order based modeling and identification of coupled creep and hysteresis effects in piezoelectric actuators. In: IEEE/asme transactions on mechatronics. https://doi.org/10.1109/TMECH.2020.2974978

23. Podlubny I (1999) Fractional differential equations. Academic Press
24. Podlubny I (2002) Geometric and physical interpretation of fractional integration and fractional differentiation. Fract Calc Appl Anal
25. Diethelm K (2010) The analysis of fractional differential equations, an application oriented, exposition using differential operators of Caputo type. Springer, Berlin
26. Li C, Ma Y (2013) Fractional dynamical system and its linearization theorem. Nonlinear Dyn 71:621–633
27. Wolfram Research, Inc., Wolfram Alpha Knowledgebase (2022) Champaign, IL, United States
28. The MathWorks Inc., MATLAB version: 9.13.0 (R2022b) (2022) Natick, Massachusetts, United States

Chapter 7
Effect of Stochastic Resetting on the Fractional Black-Scholes Equation

Guillermo Chacón-Acosta and David Del Valle

7.1 Introduction

In 1973, Fischer Black and Myron Scholes introduced an equation for pricing European financial options [1], based on asset prices that depend on market volatility and the risk-free interest rate. Their model is built under certain assumptions: at the expiration time, the option will have the strike price; assets do not pay yields; there are no transaction costs and no arbitrage, hence no risk. The dynamics of the asset price follow a geometric Brownian motion.

However, evidence suggests that the dynamics of assets do not precisely follow a geometric Brownian motion, motivating the introduction of memory in the underlying random process, similar to what happens in the continuous time random walker (CTRW). The generalization of the Black-Scholes model incorporating subdiffusive geometric Brownian motion to model asset prices was first proposed by Magdziarz in [2]. He shows that there is no single fair price for financial derivatives in such a market, i.e., the market is incomplete. This model has also been studied for baskets in [3], where it is observed that this dynamic captures the behavior of a market with few transactions, allowing the price of the asset to remain invariant during some periods, similar to a trapped test particle as it diffuses. This phenomenon has been described through Fokker-Planck equations with fractional derivatives [4]. Subsequently, the general subordinate case was studied by Li in [5] and extended to the generalized case in [6], where the moments are also calculated. It is worth mentioning that, although in this work we will review this version of the Black-Scholes equation with

G. Chacón-Acosta (✉)
Departamento de Matemáticas Aplicadas y Sistemas, Universidad Autónoma Metropolitana-Cuajimalpa, Mexico City, Mexico
e-mail: gchacon@cua.uam.mx

D. Del Valle
Licenciatura en Matemáticas Aplicadas, Universidad Autónoma Metropolitana-Cuajimalpa, Mexico City, Mexico

J. J. Rosales García et al. (eds.), *Advances in Fractional Calculus*,
Modeling and Optimization in Science and Technologies 21,
https://doi.org/10.1007/978-3-031-83989-4_7

fractional time derivative, many other variants and solution methods, both analytical and numerical, are very extensive, as detailed in [7].

On the other hand, there has been a recent interest in stochastic resetting processes, mainly used in random searches where the time to reach a target is crucial for the system [8, 9]. The resetting consists of a standard random dynamics phase, followed by a sudden interruption, bringing the dynamics back to a given initial position [10–12]. This process is randomly renewed and has been shown to be able to optimize arrival times as a function of a characteristic parameter known as the resetting rate [13]. The study of geometric Brownian motion under stochastic resetting has been performed by Stojkoski et al. in [14], where the distributions, the non-equilibrium steady state, and the corresponding moments were obtained. At long times, this system presents three regimes characterized by time intervals: frozen, unstable, and annealed steady state. To some extent, these regimes had already been explored in studies on income inequality [15, 16], where the restart acts as a stabilizing factor [17]. A renewal-like dynamics is also observed in investment portfolios with the addition or deletion of participants [18].

This chapter analyzes the effects of including a stochastic resetting process in the subdiffusive geometric Brownian motion. Specifically, we study the impact of the resetting rate on the corresponding fractional Black-Scholes equation. To this end, in the second section, we review the formulation of the Black-Scholes equation for the price of European options and its solution in the Laplace space. The subdiffusive geometric Brownian process is also analyzed through the subordination method, which gives rise to the time fractional Black-Scholes equation, whose solution is also studied in frequency space. Section three introduces the stochastic resetting process, which can also be studied using a subordination process. This section also explores the equivalences between the renewal and associated partial differential equations. By applying this analysis to the studied case, the Laplace solution of the Black-Scholes equation is obtained when a resetting is present in the underlying process. Finally, the coupling between the resetting rate and the anomalous exponent is analyzed for the case where the subdiffusive geometric Brownian process also exhibits a Poissonian-like restart. The last section discusses the results and future perspectives of the investigation.

7.2 Fractional Black-Scholes Equation and Subordination

7.2.1 Geometric Brownian Motion and Black-Scholes Formula for Option Pricing

The Black-Scholes equation applies to a process where the asset price S follows a geometric Brownian motion,

$$dS(t) = S(t)[\mu dt + \sigma dW(t)] \quad S(0) = s_0, \tag{7.1}$$

where $S(t)$ is the underlying asset price, μ is the average growth rate or expected return, σ is the volatility, and $W(t)$ is a Wiener process or standard Brownian motion [1, 19]. The probability density associated with S, i.e., the probability distribution that accounts for the risk of price fluctuations $f(s,t)$, follows the Fokker-Planck equation

$$\frac{\partial f(s,t)}{\partial t} = -\mu \frac{\partial}{\partial s}(sf(s,t)) + \frac{\sigma^2}{2}\frac{\partial^2}{\partial s^2}\left(s^2 f(s,t)\right) = \hat{L}_{FP} f(s,t), \tag{7.2}$$

which, for initial conditions $f(s,0) = \delta(s - s_0)$, follows a log-normal distribution [6],

$$f(s,t) = \frac{1}{\sqrt{2\pi\sigma^2 t}} \exp\left(\frac{(\ln s - \ln s_0 - \bar{\mu} t)^2}{2\sigma^2 t}\right), \tag{7.3}$$

where $\bar{\mu} = \mu - \sigma^2/2$. From this distribution, it follows that the mean value and mean squared displacement (MSD) have exponential dependence on time

$$\langle s \rangle = s_0 e^{\mu t}, \quad \langle s^2 \rangle = s_0^2 e^{(2\mu + \sigma^2)t}, \tag{7.4}$$

and the variance

$$\langle s^2 \rangle - \langle s \rangle = s_0^2 e^{2\mu t}\left(e^{\sigma^2 t} - 1\right). \tag{7.5}$$

From this, it is possible to calculate the price of a European option $C(S,t)$ using the CAPM financial asset pricing model, describing the relation between the asset's expected return and the risk under market equilibrium conditions [20]. Using Itô formula and calculating the expectation value, we obtain, to the lowest order in t, the well-known Black-Scholes equation

$$\frac{\partial C(s,t)}{\partial t} + \frac{\sigma^2 s^2}{2}\frac{\partial^2 C(s,t)}{\partial s^2} + rs\frac{\partial C(s,t)}{\partial s} - rC(s,t) = 0, \tag{7.6}$$

where r is the risk-free interest rate from the model. Thus, given the initial data, the expiration time T of the option, and its strike price K, it is possible to determine its cost. The final and boundary conditions can be written as follows

$$C(s,T) = \max\{s - K, 0\}, \tag{7.7}$$

$$C(0,t) = 0, \quad C(s \to \infty, t) = s, \quad t \leq T. \tag{7.8}$$

The solution is the so-called Black-Scholes formula, usually expressed as

$$C(s,t) = \frac{s}{2}\left[1 + \operatorname{erf}\left(\frac{d_+}{\sqrt{2}}\right)\right] - \frac{K}{2}e^{-r(T-t)}\left[1 + \operatorname{erf}\left(\frac{d_-}{\sqrt{2}}\right)\right], \tag{7.9}$$

where

$$d_{\pm} = \frac{\ln(s/K) + (T-t)(r \pm \sigma^2/2)}{\sigma\sqrt{T-t}}.$$

At a fixed time, option value increases as the asset price increases. As the expiration time decreases, the option value decreases to the value at expiration for a fixed asset value. Eq. (7.6) and the solution Eq. (7.9) can be interpreted as a function of the inverse or expiration time $\tau \to T - t$, such that the corresponding time derivative changes sign, and $d_{\pm}$ simplifies. We can then obtain the time Laplace transform of this expression, which requires to get the following transform of the error function that was taken from Prudnikov's book [21], Eqs. (3.7 and 3.11), i.e.,

$$\mathcal{L}\left\{\text{erf}\left(a\sqrt{\tau} \pm \frac{b}{\sqrt{\tau}}\right)\right\} = \frac{1}{\lambda}\exp\left[-2b\left(\sqrt{\lambda + a^2} \pm a\right)\right]\left(\frac{a}{\sqrt{\lambda + a^2}} \mp 1\right) \pm \frac{1}{\lambda}, \tag{7.10}$$

so that Eq. (7.9) in Laplace space is

$$\begin{aligned} C(s,\lambda) = {} & \frac{s}{\lambda}\left[1 + \frac{1}{2}\left(\frac{s}{K}\right)^{-M_+(\lambda)}\left(\frac{1}{\frac{\sigma^2}{r+\sigma^2/2}M_+(\lambda) - 1} - 1\right)\right] \\ & - \frac{K}{\lambda + r}\left[1 + \frac{1}{2}\left(\frac{s}{K}\right)^{-M_-(\lambda+r)}\left(\frac{1}{\frac{\sigma^2}{r-\sigma^2/2}M_-(\lambda+r) - 1} - 1\right)\right] \end{aligned} \tag{7.11}$$

where the following function was introduced

$$M_{\pm}(y) = \frac{\sqrt{2\sigma^2 y + (r \pm \sigma^2/2)^2} + (r \pm \sigma^2/2)}{\sigma^2}. \tag{7.12}$$

The behavior of the solution in Laplace space can be seen in Fig. 7.1.

7.2.2 *Subdiffusive Geometric Brownian Motion and Subordination*

Suppose that the asset price model is a non-Markovian process known as subdiffusive geometric Brownian motion, which is a process $\Sigma(t) = S(q(t))$, where $S(q)$ is the standard geometric process, whereas $T(q)$ is a so-called one-sided α-stable Lévy process, with parameter $0 < \alpha < 1$. The probability density is denoted $L_\alpha(t)$, and its characteristic function is $e^{-\lambda^\alpha q}$. The inverse process to $T(q)$, i.e., $Q(t)$, is defined as the collection of the first arrival times and is known as the inverse one-sided Lévy stable process. These processes are given as follows

$$\begin{aligned} dS(q) &= S(q)[\mu dq + \sigma dW(q)], \\ T(q) &= dY(q), \end{aligned} \tag{7.13}$$

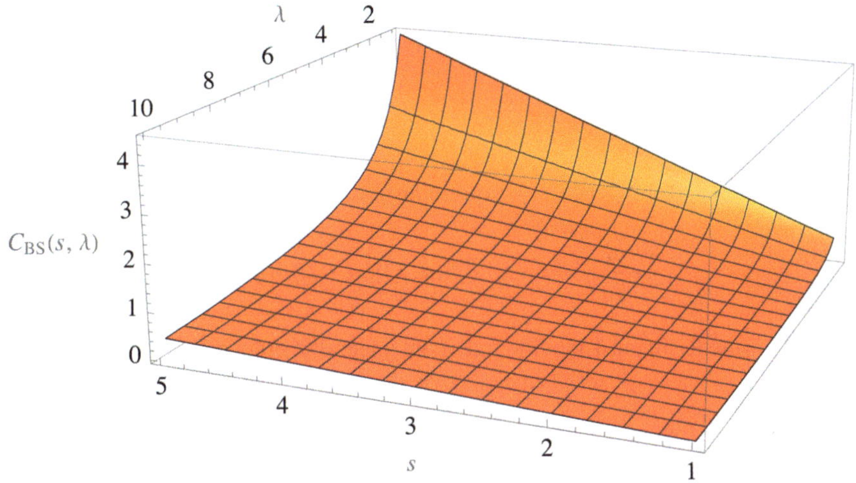

Fig. 7.1 Solution to the Black-Scholes equation in Laplace space for sample parameters $K = 1$, $r = 1$ and $\sigma = 1$

where W and Y are statistically independent. This is an extension of the geometric process that includes waiting times with a power-law distribution through the generalization given by the continuous-time random walks. There is clearly a need for a link between the time q of the standard geometric process and the real time t. Such a method of employing an auxiliary operational time related to the actual physical time is called *subordination*, [2, 5, 6, 14]. The probability density of $Q(t)$ relates the two involved times; it is called the subordination function $h(q, t)$ and is given by

$$h(q, t) := -\frac{\partial}{\partial q} \langle \Theta\left(t - T(q)\right) \rangle ,$$

where $\Theta(t)$ is the Heaviside function, and the average is calculated on T. The Laplace transform is straightforward to obtain $h(q, \lambda) = \lambda^{\alpha-1} e^{-\lambda^\alpha q}$. Then, the subordination function is the Laplace inverse of this expression which can be written in terms of Fox H–function as [5, 6]

$$h(q, t) = \frac{t}{\alpha q^{1+1/\alpha}} L_\alpha \left(\frac{t}{q^{1/\alpha}}\right) = \frac{1}{t^\alpha} H^{1,0}_{1,1} \left[\frac{q}{t^\alpha} \middle| \begin{matrix} (1-\alpha, \alpha) \\ (0, 1) \end{matrix} \right]. \tag{7.14}$$

The probability density of the process Σ is related to the density of S through the subordination function as

$$f_\Sigma(s, t) = \int dq \, h(q, t) f_S(s, q). \tag{7.15}$$

We look for the corresponding equation for the probability density of the Σ process; which can be obtained from the derivative of the subordination function

$$\frac{\partial h(q,\lambda)}{\partial q} = -\lambda^{\alpha} h(q,\lambda). \tag{7.16}$$

By Laplace inverting the previous expression

$$\frac{\partial h(q,t)}{\partial q} = -\mathcal{L}\{\lambda^{\alpha} h(q,\lambda)\} = -\mathcal{I}_0^{\alpha} h(q,t), \tag{7.17}$$

where $\mathcal{I}_0^{\alpha}$ is the Riemann-Liouville fractional integral of order $\alpha > 0$, defined as [22]

$$\mathcal{I}_0^{\alpha} g(t) = \frac{1}{\Gamma(\alpha)} \int_0^t dy \frac{g(y)}{(t-y)^{1-\alpha}}. \tag{7.18}$$

Equation (7.17) can be written with use of the Riemman-Liouville fractional derivative satisfying $\mathcal{D}_t^{\mu} = \partial_t \mathcal{I}_0^{1-\mu}$, as follows

$$\mathcal{D}_t^{1-\alpha} \frac{\partial h(q,t)}{\partial q} = -\frac{\partial h(q,t)}{\partial t}. \tag{7.19}$$

Since the density $f_s(s,q)$ follows a usual geometric Brownian motion, it satisfies the standard Fokker-Planck equation Eq. (7.2), i.e., $\partial_q f_s(s,q) = \hat{L}_{FP} f_s(s,q)$. Thus, by calculating the time derivative of (7.15), we notice that on the r.h.s., the derivative only affects $h(q,t)$, so we can use (7.19) to switch to the derivative in the auxiliary time q. Integrating by parts, and replacing $\partial_q f_s$ with the Fokker-Planck operator, the equation becomes

$$\frac{\partial f_{\Sigma}(s,t)}{\partial t} = \mathcal{D}_t^{1-\alpha} \left[-\mu \frac{\partial}{\partial s} s + \frac{\sigma^2}{2} \frac{\partial^2}{\partial s^2} s^2 \right] f_{\Sigma}(s,t). \tag{7.20}$$

The solution can be obtained by substituting the log-normal (7.3) into Eq. (7.15) with $h(q,t)$ given in Eq. (7.14), [6]. The corresponding moments are in terms of the Mittag-Leffler functions [22]

$$\langle s \rangle_{\Sigma} = s_0 E_{\alpha}(\mu t^{\alpha}), \quad \langle s^2 \rangle_{\Sigma} = s_0^2 E_{\alpha}(2\mu + \sigma^2 t^{\alpha}), \tag{7.21}$$

with the Mittag-Leffler function defined as

$$E_{\alpha}(z) = \sum_{k=0}^{\infty} \frac{z^k}{\Gamma(\alpha k + 1)}. \tag{7.22}$$

Expressions such as Eq. (7.20) can be generalized by introducing a general memory kernel instead of the power law in the waiting times leading to the definition

(7.18), [5, 6]. If the asset price follows a non-Markov process, different situations can occur depending on whether price information at different times is available at the beginning of the trading. Here, we will consider only the case when initial data and expiration date T determine the option value. In the non-Markovian case, it is also essential to consider how the trading dates will be fixed, if with the real time t or with a subordinate auxiliary time, which may lead to additional contributions.

For subdiffusive geometric Brownian motion (7.13), the option price can be obtained using the subordination function and the fact that the process $S(q)$ leads to an option whose price follows the standard Black-Scholes equation (7.6) in the following way

$$C_{fBS}(s,t) = \int dq\, h(q,t) C_{BS}(s,q), \tag{7.23}$$

where $C_{BS}(s,q)$ the solution (7.9) of Eq. (7.6). By calculating the time derivative on both sides, we find that, on the r.h.s., only the subordination function is affected again, and, by using (7.19), we can integrate by parts and solve for the temporal derivative of (7.6), thus obtaining

$$-\frac{\partial C_{fBS}(s,t)}{\partial t} = \mathcal{D}_t^{1-\alpha}\left[\frac{\sigma^2 s^2}{2}\frac{\partial^2}{\partial s^2} + rs\frac{\partial}{\partial s} - r\right] C_{fBS}(s,t) = \frac{\partial C_{fBS}(s,\tau)}{\partial \tau}. \tag{7.24}$$

The last equality occurs when considering forward options, i.e., $\tau = T - t$. In a standard Markov process, working with backward or forward time makes no difference; however, in the subdiffusive case, the standard reverse case presents an additional dependence on the initial time. Therefore, we will only consider the forward options given by Eq. (7.24). The solution can be obtained via (7.23) with the subordination function in the Laplace space [5, 6], $h(q,\lambda) = \lambda^{\alpha-1} e^{-\lambda^\alpha q}$, leading to $C_{fBS}(s,\lambda) = \lambda^{\alpha-1} C_{BS}(s,\lambda^\alpha)$, where C_{BS} is the Black-Scholes solution given in Eq. (7.11). The variation with the anomalous exponent is shown in Fig. 7.2, as α decreases the value C_{fBS} increases for large λ.

7.3 Stochastic Resetting in the Black-Scholes Model

7.3.1 *Stochastic Resetting and Subordination Scheme*

In random search processes, arbitrarily restarting the motion to a given point, e.g., to the origin, can optimize the time at which the target is reached [9, 11, 13]. This situation is known as stochastic resetting and has been the subject of study over the last decade, yielding interesting results, such as having a non-equilibrium steady state and introducing the resetting rate as a parameter in which the mean first passage time is minimized [10–12]. The idea is to start with a process that randomly explores the space, e.g., Brownian motion, and then return to the starting point after a random time whose distribution contains a resetting rate (usually the Poisson distribution) [11, 12].

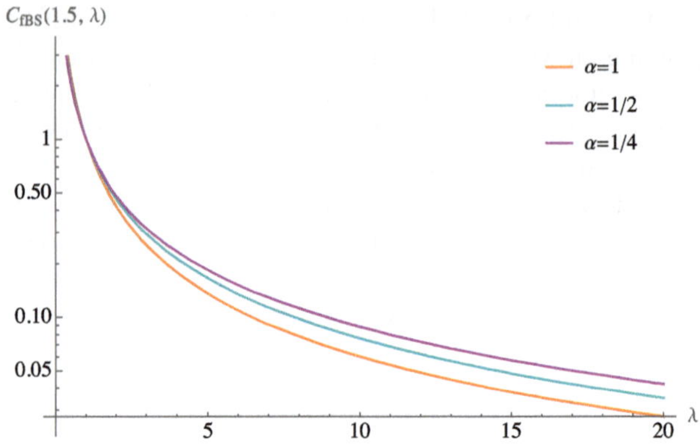

Fig. 7.2 Comparison of the solution of the fractional Black-Scholes Eq. in Laplace space for different values of α, when $s = 1.5$ and with sample parameters $K = 1$, $r = 1$ and $\sigma = 1$

Then, the equation for probability density is the equation for the underlying process plus a loss of probability at a rate ϱ at all points and a gain in probability at the restarting point that may coincide with the initial position $s = s_0$. This process can be expressed as follows

$$\frac{\partial f(s,t)}{\partial t} = \hat{L} f(s,t) - \varrho f(s,t) + \varrho \delta(s - s_0), \tag{7.25}$$

where the differential operator $\hat{L}$ stands for the Fokker-Planck equation associated with the underlying process; in the case of geometric Brownian motion it is Eq. (7.2), while in the subdiffusive case is Eq. (7.20)

One can easily realize that with the scaling $f(s,t) = e^{-\varrho t} u(s,t)$, Eq. (7.25) can be written as an equation with sources

$$\frac{\partial u(s,t)}{\partial t} - \hat{L} u(s,t) = \varrho e^{\varrho t} \delta(s - s_0), \tag{7.26}$$

whose Laplace transform is

$$\lambda u(s,\lambda) - u(s,0) - \hat{L} u(s,\lambda) = \frac{\varrho}{\lambda - \varrho} \delta(s - s_0). \tag{7.27}$$

Noticing that $u(s,0) = \delta(s - s_0)$, that the Laplace transform of $u(s,t)$ is a translation $f(s, \lambda - \varrho)$, and introducing $G_0(s,\lambda) = (\lambda - \varrho) f(s, \lambda - \varrho)/\lambda$, Eq. (7.27) becomes

$$\lambda G_0(s,\lambda) - \delta(s - s_0) = \hat{L} G_0(s,\lambda), \quad \underbrace{\Rightarrow}_{\mathcal{L}^{-1}} \quad \frac{\partial G_0(s,t)}{\partial t} = \hat{L} G_0(s,t), \tag{7.28}$$

which is the Laplace transform of the source-free equation for $G_0(s,\lambda)$. On the other hand, by changing $\lambda - \varrho \to \lambda'$, in the definition of $G_0(s,\lambda)$ and solving for $f(s,\lambda')$, we obtain

$$f(s,\lambda') = G_0(s,\lambda'+\varrho) + \frac{\varrho}{\lambda'} G_0(s,\lambda'+\varrho), \tag{7.29}$$

whose Laplace inverse is

$$f(s,t) = e^{-\varrho t} G_0(s,t) + \varrho \int_0^t d\tau\, e^{-\varrho\tau} G_0(s,\tau). \tag{7.30}$$

Equation (7.30) is the well-known renewal equation for the stochastic resetting process, and it gives the solution of Eq. (7.25), [11, 12].

The system follows the dynamics given by (7.28) until it randomly resets to the initial position, initiating a new process. Such renewal occurs at a rate ϱ. Certainly, Eq. (7.30) can be obtained through the subordination method using the following subordination function in the analog of Eq. (7.15) for this process [14],

$$h_\varrho(\eta,t) = e^{-\varrho t}\delta(t-\eta) + \varrho e^{-\varrho\eta}\Theta(t-\eta), \tag{7.31}$$

where $\Theta(t)$ is the Heaviside step function. Following this approach, it will be possible to apply the subordinations to include both the resetting and the fractional behavior in the Black-Scholes equation.

Using Eq. (7.29), with the log-nomal distribution, we can find the corresponding solution of GBM with resetting. Particularly the steady state is found in Laplace space in [14] as the limit $\lambda \to 0$, i.e.,

$$f_\varrho^{SS}(s) = \varrho f_\varrho(s,0;\varrho) \sim s^{-\gamma-1},$$

with

$$\gamma = \frac{-\bar{\mu} + \sqrt{\bar{\mu}^2 + 2\varrho\sigma^2}}{\sigma^2}.$$

and the n–th moment is

$$\langle s^n \rangle = \frac{s_0^n}{n\mu + n(n-1)\frac{\sigma^2}{2} - \varrho}\left(\mu e^{n\mu + n(n-1)\frac{\sigma^2}{2} - \varrho}\right).$$

As shown in [14], there are 3 regimes for the evolution of the moments that can be identified based on the relation between the drift, noise amplitude, and the resetting rate. First, when $\varrho > \varrho_n = n\mu + n(n-1)\sigma^2/2$, the n–th moment converges to a limiting value $\varrho/(\varrho - \varrho_n)$. At $\varrho = \varrho_n$, a sharp transition occurs, and the moment diverges linearly in time, $\sim 1 + \varrho t$. For $\varrho < \varrho_n$, this divergence becomes exponential.

7.3.2 Stochastic Resetting in Option Pricing

7.3.2.1 Black-Scholes Equation with Resetting

Considering the resetting process through subordination, it is possible to quickly obtain the governing equation for the option price $C_\varrho(s,t)$, if the underlying process has a subordination function (7.31) and by employing the corresponding Eq. (7.23), i.e.,

$$C_\varrho(s,t) = \int d\eta\, h_\varrho(\eta,t) C_{BS}(s,\eta). \tag{7.32}$$

Next, the time derivative of $C_\varrho(s,t)$, is calculated. It only affects $h_\varrho(\eta,t)$, on the r.h.s., which can be cast in terms of the η derivative, the resetting rate, and the initial conditions. The η derivative of $C_{BS}(s,\eta)$, can be replaced using the Black-Scholes equation, resulting in a similar equation to Eq. (7.25), but for the price $C_\varrho(s,t)$,

$$\frac{\partial C_\varrho(s,t)}{\partial t} = -\left[\frac{\sigma^2 s^2}{2}\frac{\partial^2}{\partial s^2} + rs\frac{\partial}{\partial s} - r\right] C_\varrho(s,t) - \varrho C_\varrho(s,t) + \varrho C_{BS}(s,0). \tag{7.33}$$

As seen in the previous section, any equation written in terms of the renewal equation admits a solution in the Laplace space given by the analog of Eq. (7.29), which, in this case, is

$$C_\varrho(s,\lambda) = \frac{\varrho+\lambda}{\lambda} C_{BS}(s,\varrho+\lambda), \tag{7.34}$$

hence, we observe the influence of resetting the Black-Scholes formula in the Laplace space (7.11). Indeed, the effect of the resetting rate on the solution of the Black-Scholes equation is to decrease C_ϱ for small λ, which is observed in Fig. 7.3.

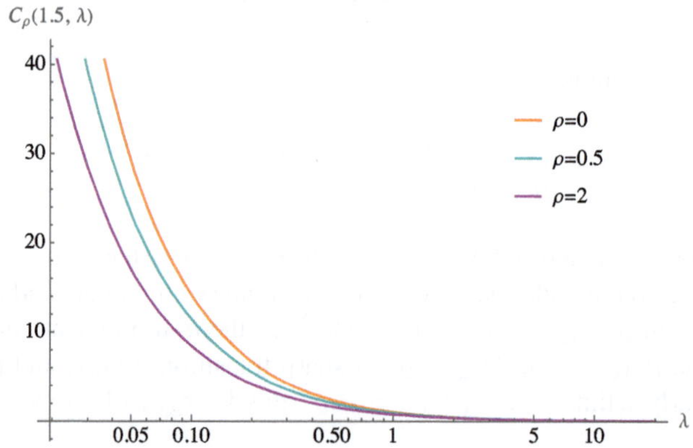

Fig. 7.3 Reset rate effect on the price C_ϱ for different values of ϱ, when $s = 1.5$ and with sample parameters $K = 1, r = 1$ and $\sigma = 1$

It is known that [22] the limit $\lim_{\to 0} C_\varrho(s,\tau)$, corresponding to expiration time, can be obtained as $\lim_{\lambda\to\infty} \lambda C_\varrho(s,\lambda)$, which by using (7.11) in (7.34), reduces to $s-K$, i.e., the resetting rate has no effect at the expiration time. On the other hand, it is well-known that a non-equilibrium steady state can be derived for the probability density for geometric Brownian motion depending on the resetting rate. Here, the limit $\tau\to\infty$ may correspond to very long expiration times, i.e.,

$$C_\varrho^{NESS}(s) := \lim_{\tau\to\infty} C_\varrho(s,\tau) = \lim_{\lambda\to 0} \lambda C_\varrho(s,\lambda) = \varrho C_{BS}(s,\varrho),$$

using (7.11) the resetting rate couples to the risk-free interest rate r and the volatility σ. Figure 7.4 shows that the effect of the resetting rate is more pronounced for small s values for this limit.

7.3.2.2 Fractional Black-Scholes Equation with Resetting

Finally, the fractional case with resetting can be handled directly in Laplace space, starting with Eq. (7.24) and subordinating the resetting time, that leads to

$$C_{\varrho fBS}(s,\lambda) = \frac{(\varrho+\lambda)^\alpha}{\lambda} C_{BS}\left(s,(\varrho+\lambda)^\alpha\right). \tag{7.35}$$

The behavior of $C_{\varrho fBS}$ can be seen in Fig. 7.5 for the case $\varrho = 2$. In this case, the fractional exponent's effect is noticed for all values of λ, and it is more pronounced for large values of s.

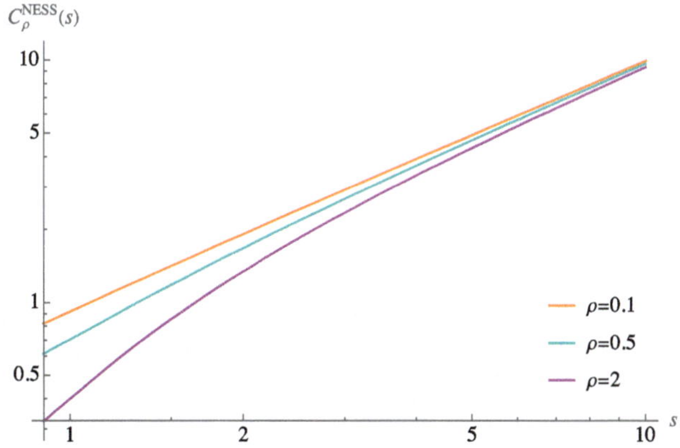

Fig. 7.4 Option price as a function of the asset price s, in the large τ limit, for different values of the resetting rate ϱ, with sample parameters $K = 1$, $r = 1$ and $\sigma = 1$

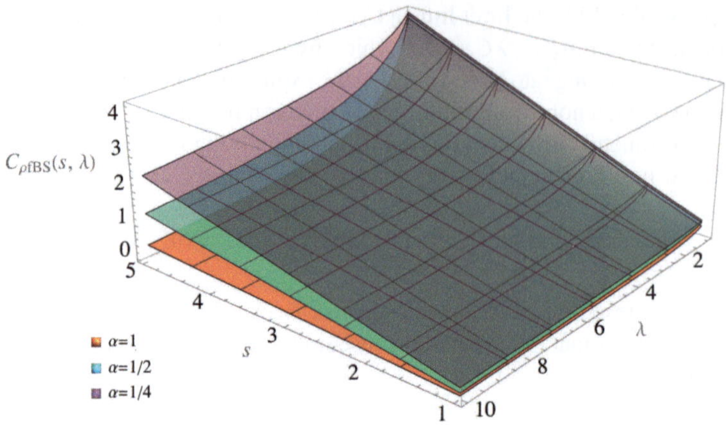

Fig. 7.5 Effect of the resetting rate fixed at $\varrho = 2$, for different values of the fractional exponent α. Same values of sample parameters $K = 1, r = 1$ and $\sigma = 1$ are used

For the maturity time the same happens as in the previous case, so that neither the reset rate nor the anomalous exponent influences the solution. However, for a large τ it is possible to obtain a solution in the stationary case where we can see the coupling of both effects.

$$C^{NESS}_{\varrho f BS}(s) = \varrho^{\alpha} C_{BS}(s, \varrho^{\alpha}),$$

which can be obtained by means of Eq. (7.11), as follows

$$\begin{aligned} C^{NESS}_{\varrho f BS}(s) = \; & s\left[1 + \frac{1}{2}\left(\frac{s}{K}\right)^{-M_{+}(\varrho^{\alpha})}\left(\frac{1}{\frac{\sigma^2}{r+\sigma^2/2}M_{+}(\varrho^{\alpha}) - 1} - 1\right)\right] \\ & - \frac{K}{1 + r\varrho^{-\alpha}}\left[1 + \frac{1}{2}\left(\frac{s}{K}\right)^{-M_{-}(\varrho^{\alpha}+r)}\left(\frac{1}{\frac{\sigma^2}{r-\sigma^2/2}M_{-}(\varrho^{\alpha}+r) - 1} - 1\right)\right], \end{aligned} \tag{7.36}$$

where the interaction between the risk-free interest rate with the resetting rate elevated to the anomalous exponent can be seen. Figure 7.6 shows this function for two different values of the resetting rate for three distinct fractional exponents. For $\varrho = 2$, $C^{NESS}_{\varrho f BS}$ increases as α decreases (solid lines), while for $\varrho = 0.2$ (dashed lines), the opposite occurs. Such behavior is more pronounced for small values of s as in the nonanomalous case.

7.4 Discussion

In this contribution, we study the Black-Scholes equation for pricing a European option under different circumstances. First, the standard Black-Scholes formula was

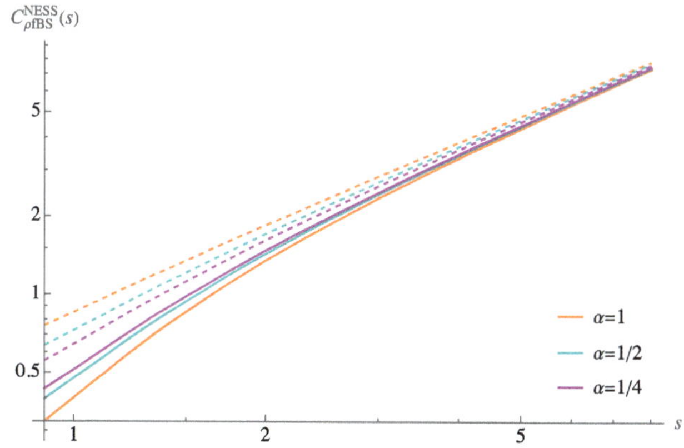

Fig. 7.6 Option price as a function of the asset price s, in the $\tau \to \infty$ limit, for different values of the fractional exponent α. The solid line corresponds to a resetting rate $\varrho = 2$, whereas dashed line is $\varrho = 0.2$, and sample parameters $K = 1$, $r = 1$ and $\sigma = 1$ are used

reviewed, depending on the asset price that follows a conventional geometric Brownian motion. The solution was expressed in Laplace space to facilitate further analysis. Then, using a time subordination process, a subdiffusive process known as the one-sided inverse Lévy process was introduced; this is achieved by a subordination function which, in ordinary time, can be described as a Fox H–function. As previously studied in the literature, there are many cases to consider when there is an underlying non-Markovian process: The order of subordination, how interest rates are taken, and how trading dates are defined significantly influence every approach. Here, we restrict ourselves to the simplest case: a time-reversed initial value problem. In this case, we obtain a fractional equation whose solution in Laplace space is relatively simple since it is the frequency raised to a fractional power. The effect of this anomalous exponent is to increase the option value at high frequencies slightly.

The so-called stochastic resetting, widely used in random search processes, was also introduced. It is usually studied from the renewal equation perspective, and can also be interpreted as a subordination process. In this case, if the underlying geometric Brownian motion exhibits a Poissonian restart with a given resetting rate ϱ, the solution of the corresponding Black-Scholes equation decreases as the resetting rate increases in the low-frequency region. The analytical Laplace inversion of the solution Eq. (7.35) is still being determined, whereas the numerical inversions without resetting process can be found in [5]. However, from the structure of (7.35), it is possible to note that some terms will have an exponential of the resetting rate competing with the interest rate and volatility. On the other hand, a property similar to the frequency shift that arises when calculating the Laplace transform of a product with an exponential is needed, but now for convolution with Mittag-Leffler functions. This, together with the fact that the inverse Laplace transform of an exponential whose argument is a power of s is related to Fox H-functions, leads us to expect that the solution is also expressible in terms of Fox H-functions.

Nevertheless, in this work, we were able to explore the asymptotic limits. Remarkably, at the expiration time limit, the same standard result is obtained irrespective of the resetting rate. On the other hand, a stationary-like case can be defined for long expiration times, so, as ϱ increases, the solution decreases. When a subdiffusive process is also considered, the resetting rate couples with the anomalous exponent similarly to the Laplace variable. As the exponent decreases, $C_{\varrho fBS}$ increases. In the stationary-like case, if the resetting rate is larger than 1, the option price increases as α decreases, while the opposite occurs for ϱ smaller than one.

More needs to be explored in this system. Some direct aspects to consider include the study of different values of the risk-free interest rate r and volatility σ, which in this analysis were taken as fixed constants, but might influence the solution by being coupled with the resetting rate and the anomalous exponent. Obtaining the inverse Laplace transform (at least some numerical approximation) will be interesting in studying the coupling effects between the resetting rate and the anomalous exponent on the system's dynamics. For instance, in [23] the Laplace perturbation homotopy method was introduced, with which the explicit solution of the model with the Caputo derivative is expressed in terms of a generalized Mittag-Leffler function, or in [24], where the reconstruction variational iteration method is used, and in [25], where even numerical methods are employed for the analysis of the subdiffusive case. These extensions to the present study will be carried out elsewhere.

Competing Interests The authors have no conflicts of interest to declare that are relevant to the content of this contribution.

Ethics Approval No ethics approval was required.

Acknowledgements We want to thank the organizers of the 6th Mexican Workshop on Fractional Calculus for inviting us to participate and for the effort to host this meeting, which is a great forum for academic exchange.

References

1. Black F, Scholes M (1973) The pricing of options and corporate liabilities. J Polit Econ 81:637–659
2. Magdziarz M (2009) Black-Scholes formula in Subdiffusive Regime. J Stat Phys 136:553–564
3. Karipova G, Magdziarz M (2017) Pricing of basket options in subdiffusive fractional Black-Scholes model. Chaos Solitons Fractals 102:245–253
4. Metzler R, Klafter J (2000) The random walk's guide to anomalous diffusion: a fractional dynamics approach. Phys Rep 339:1–77
5. Li C (2016) Option pricing with generalized continuous time random walk models. PhD Thesis, Queen Mary University of London, London, UK
6. Stojkoski V, Sandev T, Basnarkov L, Kocarev L, Metzler R (2020) Generalised geometric brownian motion: theory and applications to option pricing. Entropy 22:1432. https://doi.org/10.3390/e22121432
7. Zhang H, Zhang M, Liu F, Shen M (2024) Review of the fractional Black-Scholes equations and their solution techniques. Fractal Fract 8:101
8. Bénichou O, Loverdo C, Moreau M, Voituriez R (2011) Intermittent search strategies. Rev Mod Phys 83:1

9. Evans MR, Majumdar SN, Mallick K (2013) Optimal diffusive search: nonequilibrium resetting versus equilibrium dynamics. J Phys A Math Theor 46:185001
10. Evans MR, Majumdar SN (2011) Diffusion with stochastic resetting. Phys Rev Lett 106:160601
11. Evans MR, Majumdar SN, Schehr G (2020) Stochastic resetting and applications. J Phys A Math Theor 53:193001
12. Gupta S, Jayannavar AM (2022) Stichastic resetting: a (very) brief review. Front Phys 10:789097
13. Pal A, Prasad VV (2019) First passage under stochastic resetting in an interval. Phys Rev E 99:032123
14. Stojkoski V, Sandev T, Kocarev L, Pal A (2021) Geometric Brownian motion under stochastic resetting: a stationary yet nonergodic process. Phys Rev E 104:014121
15. Gabaix X, Lasry J-M, Lions P-L, Moll B (2016) The dynamics of inequality. Econometrica 84:2071
16. Berman Y, Peters O, Adamou A (2020) Wealth inequality and the ergodic hypothesis: evidence from the United States. J Income Distribut
17. Stojkoski V, Jolakoski P, Pal A, Sandev T, Kocarev L, Metzler R (2022) Income inequality and mobility in geometric Brownian motion with stochastic resetting: theoretical results and empirical evidence of non-ergodicity. Phil Trans R Soc A 380:20210157
18. Peters O (2010) Optimal leverage from non-ergodicity. Quant Financ 11(11):1593–1602. https://doi.org/10.1080/14697688.2010.513338
19. Hull JC (2003) Options futures and other derivatives. Pearson: Prentice Hall. NJ, USA
20. Venegas F (2008) Riesgos financieros y economicos, 2nd. Ed. Cengage Learning Eds Mexico
21. Prudnikov AP, Brychkov YA (1992) Integrals and series, vol 4. Gordon and Breach Sci. Publ, Direct Laplace Transforms
22. Sandev T, Iomin A (2023) Special functions of fractional calculus. World Scientific Publishing USA, Applications to Diffusion and Random Search Processes
23. Sawangtong P, Trachoo K, Sawangtong W, Wiwattanapataphee B (2018) The analytical solution for the black-scholes equation with two assets in the Liouville-Caputo fractional derivative sense. Mathematics 6:129. https://doi.org/10.3390/math6080129
24. Akrami MH, Poya A, Zirak MA (2024) Solving the general form of the fractional Black-Scholes with two assets through Reconstruction Variational Iteration Method. Results Appl Mathem 22:100444
25. Krzyzanowski G, Magdziarz M (2024) A tempered subdiffusive Black-Scholes model. Fract Calculus Appl Anal 27:1800–1834

Chapter 8
On the Fractional Calculus Applied to the Control of an Electrohydraulic System to Compare the Behavior in Relation to Classical PID

Edgar H. Robles, Felipe J. Torres, and Antonio Balvantin

8.1 Introduction

Hydraulic systems have a crucial role in a wide variety of industrial applications, from robotic arms, hydraulic presses to damping systems [1]. The most common hydraulic components include hydraulic cylinders for power transmission and servo valves as the system's directional control element. However, the losses in the hydraulic system could produce a non-linear behavior when considering factors such as: hydraulic fluid leakage and internal pressure variation [2].

In the control of hydraulic systems, classical PID controllers are often used because of their simple structure and practical application [3]. Nevertheless, the tuning of its three constituent parameters (integral, proportional and derivative gain) is complex for the control of systems with nonlinear behavior [4]. Therefore, several optimization tools, for example in techniques such as Particle Swarm Optimization (PSO) [5] or Genetic Algorithms (GA) [6].

A novel alternative to overcome the existence of nonlinear systems is fractional $PI^{\gamma}D^{\mu}$ controllers [4, 7], which unlike the classical PID controllers have two additional parameters in its formulation, one for the fractional integral gain (γ) and another for the fractional differential gain (μ). The advantage in the application of fractional calculus in control is its accuracy and efficiency, improving the response and stability of the system to disturbances or nonlinearities as in the case of hydraulic systems, because it can include in its model considerations such as low response

E. H. Robles (✉) · F. J. Torres · A. Balvantin
University of Guanajuato—DICIS, Salamanca, Mexico
e-mail: eh.roblessauceda@ugto.mx

F. J. Torres
e-mail: fdj.torres@ugto.mx

A. Balvantin
e-mail: antonio.balvantin@ugto.mx

J. J. Rosales García et al. (eds.), *Advances in Fractional Calculus*,
Modeling and Optimization in Science and Technologies 21,
https://doi.org/10.1007/978-3-031-83989-4_8

speed and inaccuracy in the position control [4]. Besides in [8] the development of a Fractional Order Proportional-Integral-Derivative (FOPID) position control for an electrohydraulic system is presented, considering only the experimentally acquired data to formulate the fractional control by means of an approximation algorithm (Oustaloup) and an optimization technique based on an iterative feedback fitting (IFT), the study showed that the fractional controller is faster responding than the controller based on quantitative feedback theory (QFT), being less sensitive to friction, inertia and external loads. On the other hand, fractional calculus offers the possibility of modeling and simulating hydraulic systems with viscoelastic behavior as in the case of the hydro-pneumatic suspension developed by [9], where it shows a better approximation to the experimental performance with respect to the model obtained by means of traditional calculus.

The objective of this work is to implement a fractional order position controller which consists of an asymmetric double acting hydraulic cylinder together with a 4-way and 3-position solenoid servo valve, to compare its performance quantitatively by means of the ITAE value taking as a reference the classical PID controller obtained from the literature [4].

The rest of the document is organized as follows: Section 8.2 discusses the dynamic equations governing the cylinder and servo valve. Section 8.3 presents the tuning process for the fractional controllers and outlines the parameter values used for the integer-order controllers. Section 8.4 details the simulation of fractional $PI^{\gamma}D^{\mu}$ controller, utilizing MATLAB®/Simulink® to perform numerical simulations and analyze the results. Finally, Sect. 8.5 offers the conclusions drawn from the study and the key findings of the research.

8.2 Electrohydraulic System

The electrohydraulic system, illustrated in Fig. 8.1, consists of a double-acting asymmetric hydraulic cylinder ($1A$), which is controlled via a solenoid valve ($1V1$) featuring 4-ways and 3-positions (4/3), dual solenoids, a blocked central position, and a spring return mechanism. The system is powered by a hydraulic power supply unit ($0Z$) that integrates components such as a pump-motor assembly, pressure relief valve, filter, manometer, pressure gauge, and reservoir. The design of the system exhibits several nonlinear characteristics, including static fluid friction, fluid compressibility, internal leakage, and the complex flow-pressure behavior of the control valve [10]. These nonlinearities influence the system's stability, making the application of fractional-order control advantageous for enhancing the system's performance [11].

Therefore, the integer-order dynamic equations of the hydraulic cylinder and servo valve, as presented in [4], will be utilized to model the corresponding transfer by applying the Laplace transformation to the implementation of the fractional $PI^{\gamma}D^{\mu}$ controller.

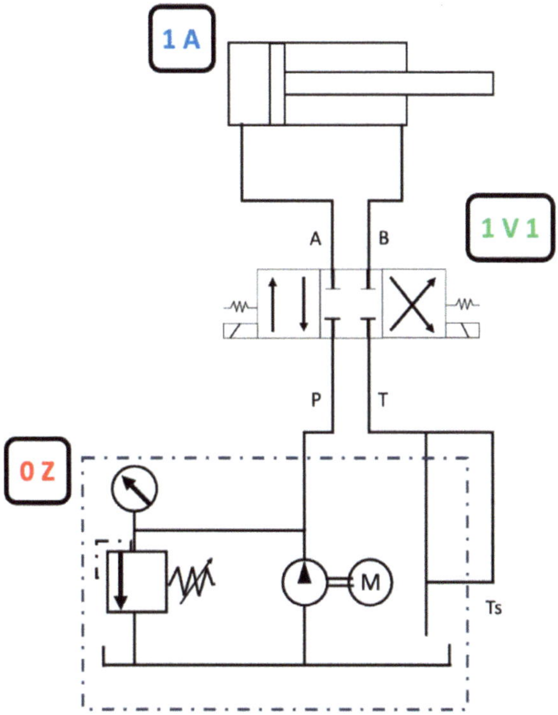

Fig. 8.1 Schematic diagram of the electro-hydraulic system

8.2.1 Hydraulic Cylinder Modeling

The dynamic model of the hydraulic cylinder in integer order, as detailed in [4], provides a classical representation of the behavior and interaction of electrohydraulic components. This model is based on three equations: a linearized equation for the hydraulic oil flow at the servo valve outlet, and two ordinary differential: one for modeling the hydraulic oil flow through the directional control device and the other addressing the force balance within the hydraulic cylinder. The model incorporates nonlinear parameters, such as the oil leakage coefficient, issues with retention seals, and variations in hydraulic oil pressure within the system.

Initially, Eq. (8.1) describes the change in flow rate in relation to the movement of the servo valve spool and the pressure in the system, where if we consider the no-load flow characteristic ($p_L = 0$), the pressure load flow will only be dependent on the internal movement of the hydraulic control device.

$$q_L = K_q x_v - K_e p_L \tag{8.1}$$

where:
q_L: system load flow rate.
K_q: flow gain on the servo valve side.
x_v: servo valve displacement.
K_e: servo valve side pressure flow.
p_L: differential pressure load.

On the other hand, the Eq. (8.2) models the hydraulic oil flow at both the inlet and outlet of the asymmetric hydraulic cylinder, taking in account the change of flow within the cylinder chamber due to the action of the piston movement, which is directly influenced by the fluid dynamics inside it. Additionally, the equations consider the flow losses which can occur due to friction or leakage and the variation of the internal pressure.

$$q_L = A_p \frac{dx_p}{dt} + C_{tp} p_L + \frac{V_t}{4\beta_e} \frac{dp_L}{dt} \tag{8.2}$$

where:
q_L: system load flow rate.
A_p: effective area of hydraulic cylinder piston.
x_p: hydraulic cylinder piston rod displacement.
C_{tp}: total leakage coefficient of hydraulic cylinder.
p_L: differential pressure load.
V_t: total volume of compression in the hydraulic cylinder chamber.
β_e: effective bulk modulus of elasticity.

Finally, the Eq. (8.3) describes the effective working provided the hydraulic cylinder as a result of the pressure differential and the effective piston area. These equations also consider factors such as the mass of the hydraulic elements, the damping force opposing motion, inertia loads and external loads acting on the system.

$$A_p\, p_L = m_t \frac{d^2 x_p}{dt^2} + B_p \frac{dx_p}{dt} + K x_p + F_L \tag{8.3}$$

where:
B_p: viscous damping factor.
K: the stiffness corresponding to the hydraulic spring.
F_L: external force applicable to the cylinder.

In order to obtain the transfer function of the hydraulic cylinder, the Laplace transformation is applied with the next considerations. The displacement of the servo valve spool as input of the system (X_v), while the displacement of the hydraulic actuator (X_p), as a output. The total flow coefficient as $K_{ce} = K_e + C_{tp}$, the ratio between $B_p << A_p^2/K_{ce}$, $K = 0$ and without external load F_L and regarding the flow $Q_L = K_q X_v$. To obtaining the following relation.

$$\frac{X_p}{Q} = \frac{\frac{1}{A_p}}{\frac{s^3}{\omega_h^2} + \frac{2\xi_h}{\omega_h}s^2 + s} \tag{8.4}$$

where:
$\omega_h = \sqrt{\frac{2\beta_e A_p}{V_t m_t}}$: natural frequency of the hydraulic cylinder.
$\xi_h = \frac{K_{ce}}{2A_p}\sqrt{\frac{2\beta_e m_t}{V_t}} + \frac{B_p}{4A_p}\sqrt{\frac{V_t}{2\beta_e\, m_t}}$, hydraulic damping ratio.

8.2.2 Servo Valve Modeling

The transfer function given by Eq. (8.4) identifies as output variable the position of the asymmetric hydraulic cylinder, and as input variable the pressure flow Q_L. Therefore, It is essential for the dynamic model of the servo valve to relate the flow rate as the output variable Q_L and the electric current I_v as the input variable. For such a formulation, in [4, 12, 13] it is recommended that the appropriate model for a transfer function of a 4-way with 3-positions should be a second-order model.

Now, considering a harmonic oscillator as a starting point for the servo valve model:

$$m_s\ddot{x}_v + f_s v\dot{x}_v + K_v x_v = K_s i_v \tag{8.5}$$

where:
m_s: servo valve mass.
f_{sv}: viscous damping coefficient.
K_v: damping constant of the servo valve spring.
K_s: gain of the servo valve.
i_v: servo valve activation current.
x_v: servo valve displacement.

Applying the Laplace transform to Eq. (8.5) and considering $Q_L = K_s X_v$, the following relation is obtained,

$$\frac{Q_L}{I_v} = \frac{K_{sf}}{\frac{s^2}{\omega_f^2} + \frac{2\xi_f}{\omega_f}s^2 + 1} \tag{8.6}$$

where:
$\frac{K_s}{m_s} = \omega_f^2$ and $\frac{f_{sv}}{m_s} = 2\varsigma_f \omega_f$.
K_{sf}: the flow gain of the servo valve.
ω_f: the intrinsic frequency of the servo valve.
ς_f: the damping ratio of the servo valve.

The transfer function defined in Eq. (8.6) characterizes the working fluid flow rate Q_L as the output and the servo valve operating current I_v as the input. This model is essential for establishing a closed-loop control system that can effectively manage the interactions between the hydraulic components.

8.3 Fractional $PI^\gamma D^\mu$ Controller

PID control has been widely employed in hydraulic systems due to its simple structure, ease of implementation, and it is one of the most used tools at an industrial [3]. The critical aspect of classical PID control involves the effective tuning of three adjustable gains: proportional, integral, and derivative gain. However, classical PID control is not adequately to control of nonlinear systems [4].

In recent decades, various intelligent optimization techniques have been utilized to overcome the challenges of controlling nonlinear and complex systems. Techniques such as Particle Swarm Optimization (PSO) [5] and Genetic Algorithms (GA) [6] have been applied for this purpose. In contrast, the PID controllers based on fractional order calculus is one of notable advantage for systems with significant nonlinearities compared to classical PID controllers [4]. However, the design of the fractional controller $PI^\gamma D^\mu$ requires the tuning of two additional parameters unlike the classical PID controller, one for the fractional integral part (γ) and another for the fractional differential part (μ). There are different tools in the literature that allow the development of the tuning of the five parameters such as: the Ziegler-Nichols method, auto tuning, selft-tuning method, particle swarm optimization [14] and optimization based on the tuning of the classical PID controller [15].

Traditionally, a fractional order controller PID is represented as follows:

$$G_c(s) = K_p + \frac{K_i}{s^\gamma} + K_d s^\mu \tag{8.7}$$

where:
K_p, K_i y K_d: integral, proportional and derivative gain.
γ: integral fractional order exponent.
μ: derivative fractional order exponent.

The application of fractional control $PI^\gamma D^\mu$ is based on the closed loop system shown in Fig. 8.2, its consider the displacement of the double-acting asymmetric hydraulic cylinder as the output variable and the current as the input variable. For

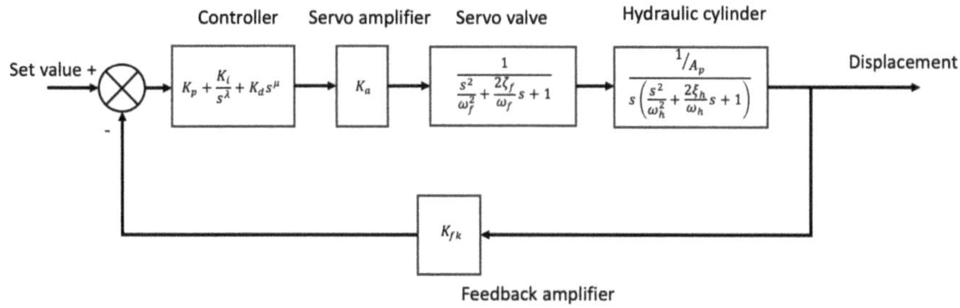

Fig. 8.2 Hydraulic servo system $PI^{\gamma}D^{\mu}$ control

tuning the additional fractional variables of the controller (γ, μ), the tuning values of the classical PID controller published in the literature were used with the MATLAB®'s *fminsearch* function and the ITAE value as the objective function.

8.3.1 *Fractional Order Controller Tuning*

For the verification of the behavior of the dynamic fractional models, performance indexes such as ISE, IAE and ITAE can be used. Being the parameter with a better acceptance the ITAE, since it considers the variation of the error by means of a temporal weighting reducing the error in the course of time, that is to say, the ITAE shows the rate at which the error approaches zero [16]. Its popularity lies in the progression of error minimization as a function of time, since parameters such as ISE or IAE consider as a criterion of the error that remains fixed in any time interval [17]. In addition, it can be found in the literature that for first-order systems the percentage overshoot or the lifting time [4] are often used as evaluation tools. For example, in [1] the application of fractional calculus to the position control of a hydraulic actuator is compared by reducing the overshoot percentage and the steady-state error value relative to classical controllers based on integer-order models. However using the ITAE value is a more complete and robust way to compare the performance of a control system by considering the evolution of the error over time [18]. It is noteworthy that the ITAE value index is applicable for systems of higher order than one with delay and without delay [19].

On the other hand, for a quantitative comparison of the performance of the classical PID controller and the fractional $PI^{\gamma}D^{\mu}$ controller, the ITAE value will be used as a suitable and commonly used metric to evaluate the performance of PID controllers [20], it is represented by the following equation:

$$ITAE = \int_0^{\infty} t\,|e(t)|\,dt \tag{8.8}$$

Table 8.1 Tuning values for controller models in fractional and integer order

Controller	PID (K_p, K_i, K_d)	$PI^{\gamma}D^{\mu}$ (γ, μ)
Classical[a]	20, 1, 0.1	1, 1
Fractional	20, 1, 0.1	0.0022, 0.0349

[a] Values taken from the literature [4]

where:
t: is the time.
$e(t)$: is the error given by the difference between the reference signal and the output signal.

To calculate the ITAE is considering a step function as an input signal for the numerical simulations, and the MATLAB® optimization function *fminsearch* to minimize the value of the ITAE by means of the 1/3 Simpson's method with fixed step. The results of the optimization process are collected in Table 8.1.

8.4 Numerical Simulations and Analysis

Simulation of fractional dynamical systems can be effectively carried out using various software platforms, including Mathematica® [21], Python® [22], and MATLAB®/Simulink® [23]. Among these options, MATLAB/Simulink stands out for its simplicity and practicality, making it the preferred choice for simulating fractional systems. Consequently, this platform will be used for the current implementation. To begin simulating fractional systems in MATLAB®/Simulink®, it is necessary to add the FOTF Toolbox developed by [24], which provides the required fractional blocks for simulating both the controller and the fractional-order transfer functions. The corresponding components are illustrated in the following Fig. 8.3.

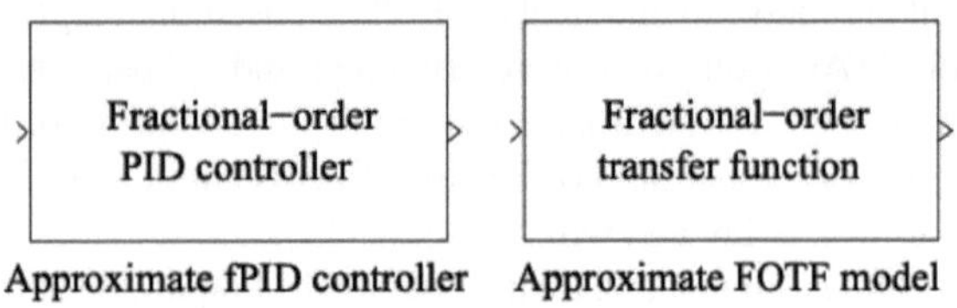

Fig. 8.3 Fractional block in Matlab/Simulink

The simulations were conducted for both the classic PID controller and the fractional-order $PI^{\gamma}D^{\mu}$ controller, as detailed in Table 8.1. These simulations were performed using the same parameters, without external disturbances. The result are presented in Fig. 8.4, which illustrates the displacement of the hydraulic cylinder in response to both step and sinusoidal input signals.

A qualitative comparison reveals that the fractional-order $PI^{\gamma}D^{\mu}$ controller enhances the system's rise time, resulting in a faster response. However, to rigorously evaluate the system's behavior, a quantitative analysis is necessary. This can be achieved by examining the Integral of time-weighted absolute error (ITAE) values, which provide a more precise measure of the system's performance. The ITAE values for both controllers are presented in Table 8.2, offering a detailed comparison of their effectiveness in improving system response.

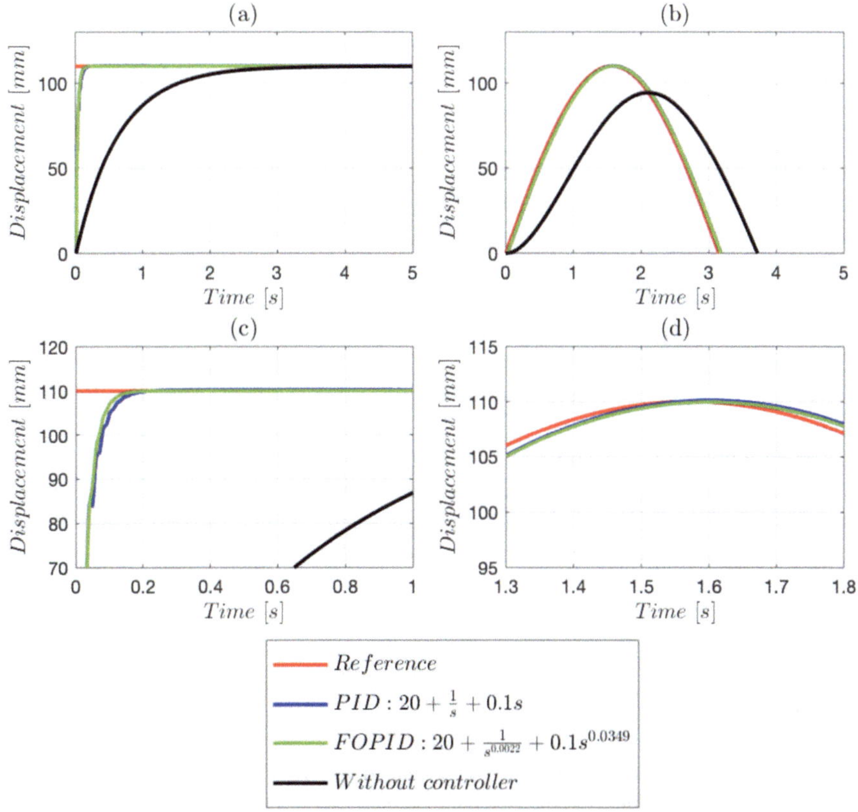

Fig. 8.4 Simulation results of the system without noise. **a** System simulation results under step function signal input; **b** System simulation under sinusoidal function input; **c** Area or interest of the results under step function signal input; **d** Area of interest of the results under sinusoidal function input

Table 8.2 ITAE values of the controllers presented without noise for step and senoidal input function

Controller	Step input	Senoidal input
Classical	0.0020	0.0248
Fractional	9.1958×10^{-5}	0.0238

Thus, the application of fractional calculus proves a advantage in systems with significant nonlinearities. This a substantial improvement is noticeable with a lower ITAE values observed across both types of input signals. These results confirm that fractional-order controllers offer superior performance in managing complex, nonlinear system dynamics compared to traditional methods.

8.5 Conclusions

In conclusion, the simulation of a fractional $PI^{\gamma}D^{\mu}$ controller in the electrohydraulic system composed by a hydraulic cylinder and a servo valve developed in MATLAB®/Simulink® has a significant improvement with respect to the classical PID controller reported in the literature. The comparison of the performance of the controllers was performed qualitatively, through the rise time and quantitatively using the ITAE performance index revealing a significant improvement in system performance under fractional order control. Thus, fractional order control is particularly useful for systems with high nonlinearities, such as the electrohydraulic model studied.

For future work, it is proposed to investigate the system response under fractional modeling, which could offer an even more accurate and efficient approach to control. In addition to instrument an electrohydraulic system to obtain experimental data to compare the performance of the fractional $PI^{\gamma}D^{\mu}$ controller with respect to its integer alternative.

References

1. Atsari ED, Halim A (2021) Design of a fractional order PID controller for electric hydraulic actuator. Kinetik: game technology, information system, computer network, computing, electronics, and control, pp 59–68
2. Gao B, Shen W, Zheng L, Zhang W, Zhao H (2022) A review of key technologies for friction nonlinearity in an electro-hydraulic servo system. Machines 10(7):568
3. Li L, Zhang X, Qian C, Wang R, Zhao M (2024) Autopilot controller of fixed-wing planes based on curriculum reinforcement learning scheduled by adaptive learning curve. IEEE Trans Emerg Top Comput Intell

4. Guo Y-Q, Zha X-M, Shen Y-Y, Wang Y-N, Chen G (2022) Research on PID position control of a hydraulic servo system based on Kalman genetic optimization. Actuators 11(6):162
5. Chaturvedi S, Kumar N, Kumar R (2024) A PSO-optimized novel PID neural network model for temperature control of jacketed CSTR: design, simulation, and a comparative study. Soft Comput 28(6):4759–4773
6. Mishra DP, Raut U, Gaur AP, Swain S, Chauhan S (2023) Particle swarm optimization and genetic algorithms for PID controller tuning. IEEE, pp 189–194
7. Boulaaras S, Rehman ZU, Abdullah FA, Jan R, Abdalla M, Jan A (2023) Coronavirus dynamics, infections and preventive interventions using fractional-calculus analysis. Aims Math
8. Maddahi A, Sepehri N, Kinsner W (2019) Fractional-order control of hydraulically powered actuators: controller design and experimental validation. IEEE/ASME Trans Mech 24(2):796–807
9. Sun H, Li R, Jikang X, Funing X, Zhang B, Dong X (2021) Fractional modeling and characteristic analysis of hydro-pneumatic suspension for construction vehicles. Processes 9(8):1414
10. Li P, Li Y (2023) Research on the electro-hydraulic servo system of picking manipulator. AIP Adv 13(1)
11. Boulaaras S, Ur Rehman Z, Abdullah FA, Jan R, Abdalla M, Jan A (2023) Coronavirus dynamics, infections and preventive interventions using fractional-calculus analysis. Aims Math 8(4):8680–8701
12. Dolatkhah Takloo S, Mardani M (2020) Mechanically closed loop gearbox test rig controller. Trans Mach Intell 3(1):1–13
13. Thayer WJ (1965) Transfer functions for Moog Servovalves. Moog Tech Bull 103:1–11
14. Saif M, Singh V, Rani A, Yadav J (2024) Particle swarm optimization based fractional order PID control for the twin rotor multi-input multi-output system. IEEE, pp 292–296
15. Tomar B, Kumar N, Sreejeth M (2024) Robust control of rotary inverted pendulum using metaheuristic optimization techniques based PID and fractional order pi λ d μ controller. J Vib Eng Technol 1–20
16. Li Y, Zhang M, Yang Y (2024) PID parameter optimization of hydro-turbine speed control system based on CPG algorithm. IOS Press, pp 124–130
17. Xue D (2017) Fractional-order control systems: fundamentals and numerical implementations, vol 1. Walter de Gruyter GmbH & Co KG
18. Naghidokht A, Khanbabaei B (2024) Optimal fractional-order PID controllers design for plasma current and horizontal position control in IR-t1 tokamak based on particle-swarm optimization. J Theor Appl Phys (JTAP) 18(3):1–12
19. Singh P, Nijhawan P, Dhingra A (2021) Tuning PID controller for three interacting tanks using ITAE performance criterion
20. Sanoj KP, Dhanya Ram V (2024) Optimal PID controller tuning for multivariable unstable systems by minimizing ITAE criteria. IFAC-PapersOnLine 57:391–396
21. Rosales JJ, Godínez FA, Martínez L, Chávez O (2021) Combining fractional derivatives to predict blood alcohol concentration. Fractals 29(03):2150063
22. Thorat SN, Ghadle KP, Muneshwar RA (2024) Simulation of fractional order 2d-mathematical model using α-fractional differential transform method. Contemp Math 685–697
23. Benbouhenni H, Hamza G, Oproescu M, Bizon N, Thounthong P, Colak I (2024) Application of fractional-order synergetic-proportional integral controller based on PSO algorithm to improve the output power of the wind turbine power system. Sci Rep 14(1):609
24. Xue D (2024) Fotf toolbox. https://www.mathworks.com/matlabcentral/fileexchange/60874-fotf-toolbox, 2024. MATLAB Central File Exchange. Retrieved September 26, 2024

4. [illegible] Fractional-order PID position control of hydraulic servo system based on [illegible]
5. [illegible]
6. [illegible]
7. [illegible]
8. [illegible]
9. [illegible]
10. [illegible]
11. [illegible]
12. [illegible]
13. [illegible]
14. [illegible]
15. [illegible]
16. [illegible] based on [illegible]
17. [illegible] control systems: [illegible] and implementations [illegible]
18. [illegible] Optimal tuning of PID controllers [illegible] optimization. [illegible]
19. Singh [illegible] (2023) Tuning PID controller for [illegible] using [illegible] performance criterion [illegible]
20. [illegible] (2019) Optimal PID controller tuning for multivariable [illegible] system by minimizing [illegible]
21. [illegible] (2021) Comparing fractional [illegible]
22. [illegible]
23. [illegible] the [illegible] power system. [illegible]
24. [illegible]

Chapter 9
Bandwidth of Matrix Operators Raised to a Fractional Power

Jorge Ramos-Grez and Mihir Sen

9.1 Motivation

It is common to design an undergraduate mechanical engineering curriculum with a heavy dose of mathematics, normally including differential and integral calculus, and linear algebra. This leads to differential equations which model the machines and processes that are studied in upper-level courses. The entire sequence of courses is presented in such a manner that continuous functions represent displacement and other physical variables which can then be measured to verify the mathematical model. However, this sequence should be the other way around: *i.e.* experimentation should come first and this would necessarily lead to mathematical modeling.

Measured data can be represented as a matrix vector $\mathbf{x} = (x_1, x_2, \ldots, x_n)$. To a system, we have thus an input $x_{in}(t)$ and an output $x_{out}(t)$, where t is time, both represented as matrices. The relation between x_{in} and x_{out} is a mathematical model. One wishes to have this mathematical model for design, control, or for any other similar situation in which we do not have the actual machine to work with but still have to make predictions as to how it will behave. If we have the actual machine, we can simply make measurements on it. Though usually both space and time enter into a physical problem, we will look at only the temporal development of the system variable as a function of time, and ignore its spatial dependence for now. The spatial dependence will be a simple extension.

Fractional-order derivatives have been shown to play a role in the modeling of actual systems [1, 2]. There are many applications of fractional-order equations and

J. Ramos-Grez (✉)
Pontificia Universidad Católica de Chile, School of Engineering, Santiago, Chile
e-mail: jramos@uc.cl

M. Sen
University of Notre Dame, College of Engineering, Notre Dame, USA
e-mail: Mihir.Sen.1@nd.edu

J. J. Rosales García et al. (eds.), *Advances in Fractional Calculus*,
Modeling and Optimization in Science and Technologies 21,
https://doi.org/10.1007/978-3-031-83989-4_9

there are many mathematical models of physical systems based on fractional-order derivatives [3–5].

9.2 Introduction

The first order forward in time derivative operator **D** acting on a function **x(t)** with $O(1)$ approximation can be expressed as,

$$\mathbf{D}\mathbf{x}(t) = \frac{d\mathbf{x}(t)}{dt} = \frac{x_i - x_{i-1}}{\Delta t} + O(1) \tag{9.1}$$

with i a positive integer index such that $i = 0, 1, 2....$, and $t = i\Delta t$

In this same manner, one can consider a second-order derivative operator $\mathbf{D}^2$ such that,

$$\mathbf{D}\mathbf{D}\mathbf{x}(t) = \frac{d}{dt}\frac{d\mathbf{x}(t)}{dt} = \frac{d\mathbf{x}^2(t)}{dt^2} = \frac{d}{dt}\frac{x_i - x_{i-1}}{\Delta t} \tag{9.2}$$

$$= \frac{(x_{i+1} - x_i) - (x_i - x_{i-1})}{\Delta t^2} = \frac{x_{i+1} - 2x_i + x_{i-1}}{\Delta t^2} + O(2) \tag{9.3}$$

which corresponds to a second-order finite central difference in time.

Analogously, an operator **D**, can also be expressed in terms of the product of its semi-derivative operator $\mathbf{D}^{\frac{1}{2}}$, such that,

$$\mathbf{D} = \mathbf{D}^{\frac{1}{2}}\mathbf{D}^{\frac{1}{2}} \tag{9.4}$$

Generalizing to any real fractional power α, we can say that,

$$\mathbf{D}^{\alpha} = \frac{d^{\alpha}}{dt^{\alpha}} \tag{9.5}$$

thus, for any integer p and q, the operator **D** can be raised to $\frac{p}{q}$ power, obtaining the following fractional order derivative operator,

$$\mathbf{D}^{\frac{p}{q}} = (\mathbf{D}^{p})^{\frac{1}{q}} = (\mathbf{D}^{\frac{1}{q}})^{p} \tag{9.6}$$

which is indifferent regarding the sequence of power-raising operations. In differential form, it can be expressed as,

$$\mathbf{D}^{\frac{p}{q}}\mathbf{x}(t) = \frac{d^{\frac{1}{q}}}{dt^{\frac{1}{q}}}\frac{d^p\mathbf{x}(t)}{dt^p} = \frac{d^p}{dt^p}\frac{d^{\frac{1}{q}}\mathbf{x}(t)}{dt^{\frac{1}{q}}} \tag{9.7}$$

regardless of the order of differentiation. In the same manner,

$$\mathbf{D}^{p+\frac{1}{q}} = \mathbf{D}^p\mathbf{D}^{\frac{1}{q}} = \mathbf{D}^{\frac{1}{q}}\mathbf{D}^p \tag{9.8}$$

which is indifferent regarding the sequence of operations; however, this commutative property may not necessarily hold true always as it will depend on the nature of the differential operator.

Consider now a discretized first-order derivative operator $\mathbf{D}$ given by a $n \times n$ square matrix (from now on all matrices will be square of $n \times n$; vectors will be $n \times 1$). For example, for $n = 5$ we have a matrix

$$\mathbf{D} = \frac{1}{\Delta t}\begin{pmatrix} 1 & 0 & 0 & 0 & 0 \\ -1 & 1 & 0 & 0 & 0 \\ 0 & -1 & 1 & 0 & 0 \\ 0 & 0 & -1 & 1 & 0 \\ 0 & 0 & 0 & -1 & 1 \end{pmatrix}_{n\times n} \tag{9.9}$$

where Δt is the time interval between measurements (assumed equal). Then $\mathbf{D}\mathbf{x}(t)$ is the (time-)derivative of vector $\mathbf{x}(t)$ where

$$\mathbf{x}(t) = \begin{pmatrix} x_1(t) \\ x_2(t) \\ \vdots \\ x_n(t) \end{pmatrix}_{n\times 1} \tag{9.10}$$

The elements of the vector are from experiments and the first and last rows of $\mathbf{D}\mathbf{x}(t)$ are usually neglected. This is how time derivatives are normally obtained from experimental measurements.

The matrix $\mathbf{D}$ will be called the *derivative operator* since

$$\mathbf{D}\mathbf{x}(t) = \left(\ldots\, 0,\, x_i - x_{i-1}\, ,0 \ldots\right) \tag{9.11}$$

gives the first time-derivative of $\mathbf{x}(t)$, as long as Δt is suitably small.

The derivative operator is not unique, but we have selected one that is backward in time to not violate causality in time.

Of course, the matrix $\mathbf{D}$ can be, and usually is, larger than $n = 5$, though here we will use $n = 5$ to not take up so much space in its writing, though it is understood that any large value of n may be used. It all depends on the system, the measuring instrument used, and the sampling rate. All that is necessary for the matrix $\mathbf{D}$, of whatever size, is for the diagonal in Eq. (9.9) to be all 1's and the first diagonal below that to be -1's, the rest being zeros.

Assume that a system is linear so that its output and input will be related in the following manner

$$\mathbf{x}_{out} = \mathbf{F}(\mathbf{x}_{in}) \tag{9.12}$$

where the subscripts *out* and *in* represent the output and input of the system, respectively. Of course, the function **F** will be complicated, and will perhaps be a combination of matrices. In any case, this will then be the mathematical model that represents the system. The simplest situation arises if $\mathbf{F} = \mathbf{D}$, so that the output $x_{out}(t)$ is simply the time-derivative of the input $x_{in}(t)$.

9.3 Matrix Functions

The exponential of a scalar x is $\exp(x)$, which is just another way of writing e^x, and its natural logarithm is $\ln(x)$. It can be easily seen that $\exp(\ln(x)) = x$, i.e. the exp and log operators are inverses of each other. The series representation below of the exponential of a matrix **A** is a *definition*, while that of its logarithm is derivable from the former. Both can be used within MATLAB® (or similar software such as Julia, GNU Octave, NumPy, SciLab, Sage, MATHCAD, etc.; we used MATLAB here simply because it was conveniently available to us, as it is in many institutions), and are calculated internally as follows, where **A** is any square matrix, and **I** is the identity matrix. The two series below are finite if **A** is nilpotent.

- Exponential of a matrix

$$\exp(\mathbf{A}) = \mathbf{I} + \mathbf{A} + \frac{\mathbf{A}^2}{2!} + \frac{\mathbf{A}^3}{3!} + \dots \tag{9.13}$$

  This is a definition which is inspired by the infinite series

$$\exp(x) = 1 + x + \frac{x^2}{2!} + \frac{x^3}{3!} + \cdots, \tag{9.14}$$

  when x is a scalar.
- Natural logarithm of a matrix

$$\ln(\mathbf{I} + \mathbf{A}) = \mathbf{A} - \frac{\mathbf{A}^2}{2} + \frac{\mathbf{A}^3}{3} - \cdots \tag{9.15}$$

  This is *not* a definition but is derivable from the exp() function. In general, the logarithm is non-unique but is similar in form to the series

$$\ln(1+x) = x - \frac{x^2}{2} + \frac{x^3}{3} - \cdots \tag{9.16}$$

where x is a scalar.

MATLAB includes the commands $expm(\mathbf{A})$ and $logm(\mathbf{A})$. Each of these operations, $expm$ and $logm$, on a matrix $\mathbf{A}$ will produce a matrix.

9.4 Integer-Order Derivatives

The first-order matrix operator $\mathbf{D}$ can be raised to any power r as the derivative operator of order r. Thus $\mathbf{D}^r$ is the r-th order derivative and $\mathbf{D}^r\mathbf{x}$ is the result of its operation on vector $\mathbf{x}$. Thus higher-order derivatives can be obtained by repeated application of the first-order derivative, i.e. $\mathbf{D}^2, \mathbf{D}^3$, and so on, are derivatives of order two, three, etc. respectively.

9.4.1 Derivative of Order One

This is a time-derivative given by Eq. (9.9), such that $\mathbf{D}\mathbf{A}^q = q\mathbf{A}^{q-1}$, where $\mathbf{A} = \mathbf{A}(t)$.

9.4.2 Derivatives of Other Orders

Higher-order derivatives can then be obtained by repeated application of the first-order derivative, i.e.

$$\mathbf{D}^2\mathbf{A} = -\mathbf{D}\mathbf{D}^T\mathbf{A} = \mathbf{D}_2\mathbf{A} \tag{9.17}$$

$$\mathbf{D}^3\mathbf{A} = \mathbf{D}\mathbf{D}^T\mathbf{D}\mathbf{A} = \mathbf{D}_3\mathbf{A} \tag{9.18}$$

$$\text{etc..} \tag{9.19}$$

For example using Eq. (9.9)

$$\mathbf{D}_2 = \frac{-1}{\Delta t}\begin{pmatrix} 1 & 0 & 0 & 0 & 0 \\ -1 & 1 & 0 & 0 & 0 \\ 0 & -1 & 1 & 0 & 0 \\ 0 & 0 & -1 & 1 & 0 \\ 0 & 0 & 0 & -1 & 1 \end{pmatrix} \frac{1}{\Delta t}\begin{pmatrix} 1 & -1 & 0 & 0 & 0 \\ 0 & 1 & -1 & 0 & 0 \\ 0 & 0 & 1 & -1 & 0 \\ 0 & 0 & 0 & 1 & -1 \\ 0 & 0 & 0 & 0 & 1 \end{pmatrix} = \frac{1}{\Delta t^2}\begin{pmatrix} -1 & 1 & 0 & 0 & 0 \\ 1 & -2 & 1 & 0 & 0 \\ 0 & 1 & -2 & 1 & 0 \\ 0 & 0 & 1 & -2 & 1 \\ 0 & 0 & 0 & 1 & -2 \end{pmatrix} \tag{9.20}$$

9.5 Fractional-Order Derivatives

We propose a similar definition for $\mathbf{D}^{p/q}$, where p and q are unequal real integers, $p < q$, as a fractional-order derivative. Thus

$$\mathbf{D}^p = \mathbf{A}^q \tag{9.21}$$

where $\mathbf{A} = \mathbf{D}^{p/q}$. Hence $\mathbf{A}$ is a matrix operator representing a derivative of order p/q.

In order to compute $\mathbf{D}^{p/q}$, we take advantage of the $expm()$ and $logm()$ functions defined in MATLAB, in the following procedure,

$$\mathbf{A} = \mathbf{D}^{p/q} = expm(\frac{p}{q} logm(\mathbf{D})) \tag{9.22}$$

However, we can also use the infinite series definition of $exp()$ and $ln()$ to express a fractional matrix as a function of its exponent and matrix,

$$\mathbf{A} = \mathbf{D}^{p/q} = exp(\frac{p}{q} ln(\mathbf{D})) + O(error) \tag{9.23}$$

9.6 Bandwidth of Matrix

It is easy to visually observe in a banded, square matrix what the bandwidth is, but we must define it here in a manner that is also suitable for numerical computation on arbitrary matrices. Thus we *define* the bandwidth distance d_{Matrix} of a square matrix $\mathbf{A}$ as

$$d_{Matrix} = \left(\sum_{i,j} a_{ij}^2 \right)^{\frac{1}{2}} \tag{9.24}$$

where $a_{i,j}$ is the i,j elements of the matrix and here we are adding the square of each element, as zero elements will not contribute to the d value. Then the square root of this sum is obtained.

Moreover, a bandwidth distance, $d_{Diagonal}$, for each of the non-zero diagonals of the matrix $\mathbf{A}$ can be computed much in the same manner,

$$d_{Diagonal} = \left(\sum_{i=1:N, j=1:N+1-i} a_{i+j-1,j}^2 \right)^{\frac{1}{2}} \tag{9.25}$$

9.7 Examples

Four examples are presented below and will show the manner in which fractional-order derivatives may be used. The first two, consider several 5 time nodes, (*i.e.*, a matrix of rank 5) and fractional powers, $p/q = 1/2$ and $p/q = 2/3$. The other two examples consider fractional powers of $p/q = 1/2$ but for 6 and 7 times nodes, respectively (*i.e.*, a matrix of rank 6 and 7). These will be computed using MATLAB built-in functions and the previously presented infinite series procedure with approximations of $O(2)$ and $O(4)$, respectively.

- Considering five time nodes N = 5 (matrix of rank 5), $p = 1, q = 2$ so that $p/q = 1/2$. This is a derivative of order 1/2 (also called a semi-derivative). In MATLAB it can also be obtained using the command $sqrtm(\mathbf{D})$ [6], and it can be verified that the result is the same when calculated using MATLAB commands, $expm()$ and $logm()$, using the procedure suggested before.

$$\mathbf{D}^{\frac{1}{2}} = \mathbf{sqrtm(D)} \tag{9.26}$$

$$\mathbf{D}^{\frac{1}{2}} = \frac{1}{\Delta t^{\frac{1}{2}}} \begin{pmatrix} 1 & 0 & 0 & 0 & 0 \\ -1 & 1 & 0 & 0 & 0 \\ 0 & -1 & 1 & 0 & 0 \\ 0 & 0 & -1 & 1 & 0 \\ 0 & 0 & 0 & -1 & 1 \end{pmatrix}^{\frac{1}{2}} = \frac{1}{\Delta t^{\frac{1}{2}}} \begin{pmatrix} 1 & 0 & 0 & 0 & 0 \\ -0.5000 & 1 & 0 & 0 & 0 \\ -0.1250 & -0.5000 & 1 & 0 & 0 \\ -0.0625 & -0.1250 & -0.5000 & 1 & 0 \\ -0.0391 & -0.0625 & -0.1250 & -0.5000 & 1 \end{pmatrix} \tag{9.27}$$

vis the same as when,

$$\mathbf{D}^{\frac{1}{2}} = \mathbf{expm}(\frac{\mathbf{1}}{\mathbf{2}}\mathbf{logm(D)}) \tag{9.28}$$

and also equal, up to 4 decimal places, to the infinite series procedure with $O(4)$ approximation. In this particular case the matrix bandwidth distance d = 2.4609. However, when using a procedure with approximation $O(2)$, it renders,

$$\mathbf{D}^{\frac{1}{2}} = \frac{1}{\Delta t^{\frac{1}{2}}} \begin{pmatrix} 1 & 0 & 0 & 0 & 0 \\ -0.5000 & 1 & 0 & 0 & 0 \\ -0.1250 & -0.5000 & 1 & 0 & 0 \\ 0.1250 & -0.1250 & -0.5000 & 1 & 0 \\ 0.0312 & 0.1250 & -0.1250 & -0.5000 & 1 \end{pmatrix} \tag{9.29}$$

showing a deviation in the last two diagonals and a slightly larger matrix bandwidth distance d = 2.4656. From these results, it can be observed that when this matrix operator **D** is raised to a fractional power, the lower matrix triangle is completely populated. This could be interpreted as the fractional operator having properties of memory effect.

- Considering five times nodes N = 5 (matrix of rank 5), $p = 2, q = 3$ so that $p/q = 2/3$. This is a derivative of fractional order 2/3. Thus,

$$\mathbf{D}^{\frac{2}{3}} = \frac{1}{\Delta t^{\frac{2}{3}}} \begin{pmatrix} 1 & 0 & 0 & 0 & 0 \\ -1 & 1 & 0 & 0 & 0 \\ 0 & -1 & 1 & 0 & 0 \\ 0 & 0 & -1 & 1 & 0 \\ 0 & 0 & 0 & -1 & 1 \end{pmatrix}^{\frac{2}{3}} = \frac{1}{\Delta t^{\frac{2}{3}}} \begin{pmatrix} 1 & 0 & 0 & 0 & 0 \\ -0.6667 & 1 & 0 & 0 & 0 \\ -0.1111 & -0.6667 & 1 & 0 & 0 \\ -0.0494 & -0.1111 & -0.6667 & 1 & 0 \\ -0.0288 & -0.0494 & -0.1111 & -0.6667 & 1 \end{pmatrix} \tag{9.30}$$

Here again, the results using MATLAB commands are the same up to 4 decimal places to the proposed procedure with $O(4)$ approximation . In this particular case the matrix bandwidth distance $d = 2.6116$. However, when using a procedure with approximation $O(2)$, it renders,

$$\mathbf{D}^{\frac{2}{3}} = \frac{1}{\Delta t^{\frac{2}{3}}} \begin{pmatrix} 1 & 0 & 0 & 0 & 0 \\ -0.6667 & 1 & 0 & 0 & 0 \\ -0.1111 & -0.6667 & 1 & 0 & 0 \\ 0.2222 & -0.1111 & -0.6667 & 1 & 0 \\ 0.0556 & 0.2222 & -0.1111 & -0.6667 & 1 \end{pmatrix} \tag{9.31}$$

showing a deviation in the last two diagonals and a slightly larger matrix bandwidth distance $d = 2.630$.

- Considering now six time nodes N = 6 (matrix of rank 6), $p = 1, q = 2$ so that $p/q = 1/2$. This is a derivative of order 1/2. Thus,

$$\mathbf{D}^{\frac{1}{2}} = \mathbf{sqrtm(D)} = \frac{\mathbf{1}}{\mathbf{\Delta t^{\frac{1}{2}}}} \begin{pmatrix} 1 & 0 & 0 & 0 & 0 & 0 \\ -0.5000 & 1 & 0 & 0 & 0 & 0 \\ -0.1250 & -0.5000 & 1 & 0 & 0 & 0 \\ -0.0625 & -0.1250 & -0.5000 & 1 & 0 & 0 \\ -0.0391 & -0.0625 & -0.1250 & -0.5000 & 1 & 0 \\ -0.0273 & -0.0391 & -0.0625 & -0.1250 & -0.5000 & 1 \end{pmatrix} \tag{9.32}$$

For an $O(4)$ approximation,

$$\mathbf{D}^{\frac{1}{2}} = \frac{1}{\Delta t^{\frac{1}{2}}} \begin{pmatrix} 1 & 0 & 0 & 0 & 0 & 0 \\ -0.5000 & 1 & 0 & 0 & 0 & 0 \\ -0.1250 & -0.5000 & 1 & 0 & 0 & 0 \\ -0.0625 & -0.1250 & -0.5000 & 1 & 0 & 0 \\ -0.0391 & -0.0625 & -0.1250 & -0.5000 & 1 & 0 \\ -0.0271 & -0.0391 & -0.0625 & -0.1250 & -0.5000 & 1 \end{pmatrix} \tag{9.33}$$

There is only a slight deviation at the 4th decimal place in the last diagonal. The matrix bandwidth distance $d = 2.7070$ is the same in both cases, but it has increased considerably when compared to the 5-nodes example.
For an $O(2)$ approximation,

$$\mathbf{D}^{\frac{1}{2}} = \frac{1}{\Delta t^{\frac{1}{2}}} \begin{pmatrix} 1 & 0 & 0 & 0 & 0 & 0 \\ -0.5000 & 1 & 0 & 0 & 0 & 0 \\ -0.1250 & -0.5000 & 1 & 0 & 0 & 0 \\ 0.1250 & -0.1250 & -0.5000 & 1 & 0 & 0 \\ 0.0312 & 0.1250 & -0.1250 & -0.5000 & 1 & 0 \\ 0.0000 & 0.0312 & 0.1250 & -0.1250 & -0.5000 & 1 \end{pmatrix} \tag{9.34}$$

here we can see that the last three diagonals deviate considerably. The matrix bandwidth distance $d = 2.7132$ has increased, but it is considerably larger compared to the 5-node example.

- Considering seven time nodes N = 7 (matrix of rank 7) , $p = 1, q = 2$ so that $p/q = 1/2$. This is a derivative of order 1/2.

$$\mathbf{D}^{\frac{1}{2}} = \mathbf{sqrtm(D)} = \frac{\mathbf{1}}{\mathbf{\Delta t^{\frac{1}{2}}}} \begin{pmatrix} 1 & 0 & 0 & 0 & 0 & 0 & 0 \\ -0.5000 & 1 & 0 & 0 & 0 & 0 & 0 \\ -0.1250 & -0.5000 & 1 & 0 & 0 & 0 & 0 \\ -0.0625 & -0.1250 & -0.5000 & 1 & 0 & 0 & 0 \\ -0.0391 & -0.0625 & -0.1250 & -0.5000 & 1 & 0 & 0 \\ -0.0273 & -0.0391 & -0.0625 & -0.1250 & -0.5000 & 1 & 0 \\ -0.0205 & -0.0273 & -0.0391 & -0.0625 & -0.1250 & -0.5000 & 1 \end{pmatrix} \tag{9.35}$$

With an $O(4)$ approximation,

$$\mathbf{D}^{\frac{1}{2}} = \frac{1}{\Delta t^{\frac{1}{2}}} \begin{pmatrix} 1 & 0 & 0 & 0 & 0 & 0 & 0 \\ -0.5000 & 1 & 0 & 0 & 0 & 0 & 0 \\ -0.1250 & -0.5000 & 1 & 0 & 0 & 0 & 0 \\ -0.0625 & -0.1250 & -0.5000 & 1 & 0 & 0 & 0 \\ -0.0391 & -0.0625 & -0.1250 & -0.5000 & 1 & 0 & 0 \\ -0.0271 & -0.0391 & -0.0625 & -0.1250 & -0.5000 & 1 & 0 \\ -0.0635 & -0.0271 & -0.0391 & -0.0625 & -0.1250 & -0.5000 & 1 \end{pmatrix} \tag{9.36}$$

There is a slight deviation at the 4th decimal place in the penultimate diagonal and a larger deviation at the 2nd decimal place in the last diagonal. The matrix bandwidth distance $d = 2.9332$ is almost the same in both cases, and it has increased considerably when compared to the matrix of rank 5.
For an $O(2)$ approximation,

$$\mathbf{D}^{\frac{1}{2}} = \frac{1}{\Delta t^{\frac{1}{2}}} \begin{pmatrix} 1 & 0 & 0 & 0 & 0 & 0 & 0 \\ -0.5000 & 1 & 0 & 0 & 0 & 0 & 0 \\ -0.1250 & -0.5000 & 1 & 0 & 0 & 0 & 0 \\ 0.1250 & -0.1250 & -0.5000 & 1 & 0 & 0 & 0 \\ 0.0312 & 0.1250 & -0.1250 & -0.5000 & 1 & 0 & 0 \\ 0.0000 & 0.0312 & 0.1250 & -0.1250 & -0.5000 & 1 & 0 \\ 0.0000 & 0.0000 & 0.0312 & 0.1250 & -0.1250 & -0.5000 & 1\ 0 \end{pmatrix} \tag{9.37}$$

here we can see that the last three diagonals deviate considerably. The matrix bandwidth distance $d = 2.9400$ has increased, and it is considerably larger compared to the matrix rank 5 cases.

Figures 9.1 and 9.2 illustrate the bandwidth distance, d, of **D** matrices constructed varying the matrix rank from 4 up to 10 and the fractional power from $p/q = 1/2$, $2/3$, $3/4$, $4/5$ and $5/6$. From these, we can observe that as the rank of the matrix increases the d value also increases monotonously, however showing decreasing increments behavior. Additionally, as the fractional power p/q increases to 1.0 the values of the bandwidth distance, d also increase. In Fig. 9.3 it can be verified that the bandwidth distance computed with approximation O(2) is always larger than when computed with a higher order approximation O(4) for a power $p/q = 0.80$. Finally, in Fig. 9.4 the bandwidth distance values of each none-zero diagonal of a **D** matrix of rank 10 are computed under approximation $O(4)$ and $O(2)$ for a power of $p/q = 0.8333$. From the latter it can be observed the decay in the d value as the diagonal is further away from the main diagonal and that the distance values computed under approximation $O(2)$ and $O(4)$ are the same for the first three diagonal and for further diagonals their relative values oscillate. Additionally, only the first 6 diagonals to the left from the main diagonal are non-zero, indicating that a maximum diagonal bandwidth of 7 appears in differential forward in time operator raised to a power less than 1.

9.8 Case Study: Fractional Derivative of an Exponential Function

Consider the fractional derivative of order $\frac{1}{2}$ in the variable t of the exponential function $\mathbf{e}^t$, obtained from a power series approximation scheme,

$$\frac{\mathbf{d}^{0.5}}{\mathbf{d}t^{0.5}}\mathbf{e}^t = t^{-0.5} \sum_{k=0}^{\infty} \frac{t^k}{\Gamma(k - \frac{1}{2} + 1)} \tag{9.38}$$

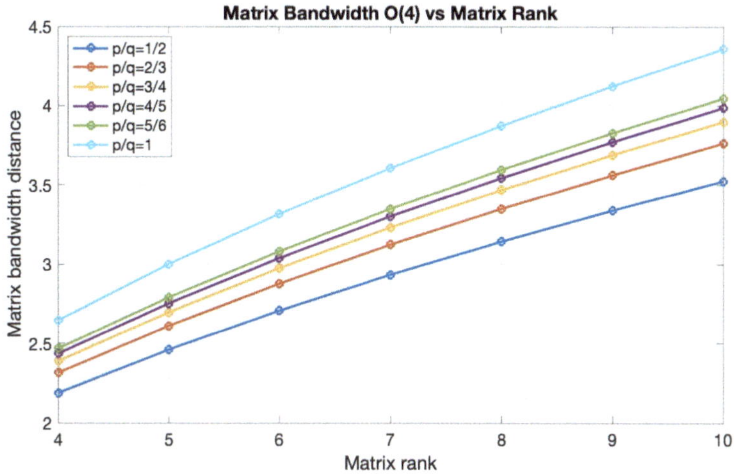

Fig. 9.1 Matrix bandwidth distance versus matrix rank for different fractional powers of the operator matrix and approximation orders. $O(4)$ approximation and p/q powers from 1/2–5/6

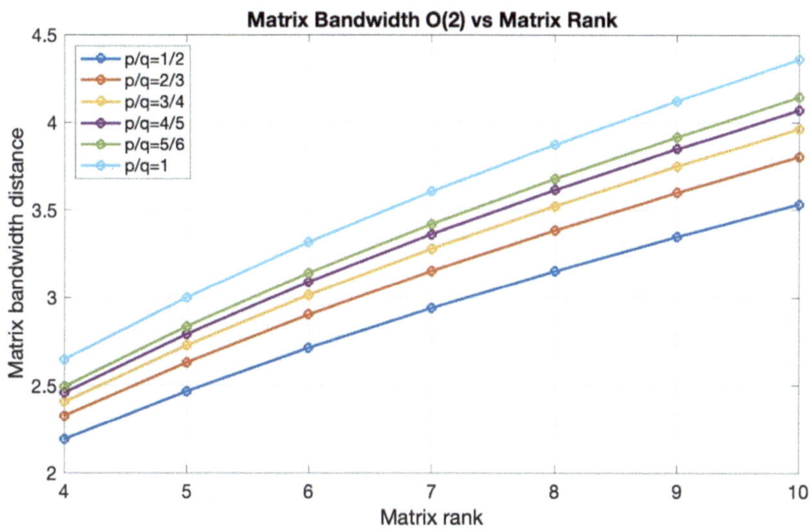

Fig. 9.2 Matrix bandwidth distance versus matrix rank for different fractional powers of the operator matrix and approximation orders. O(2) approximation and p/q powers from 1/2–5/6

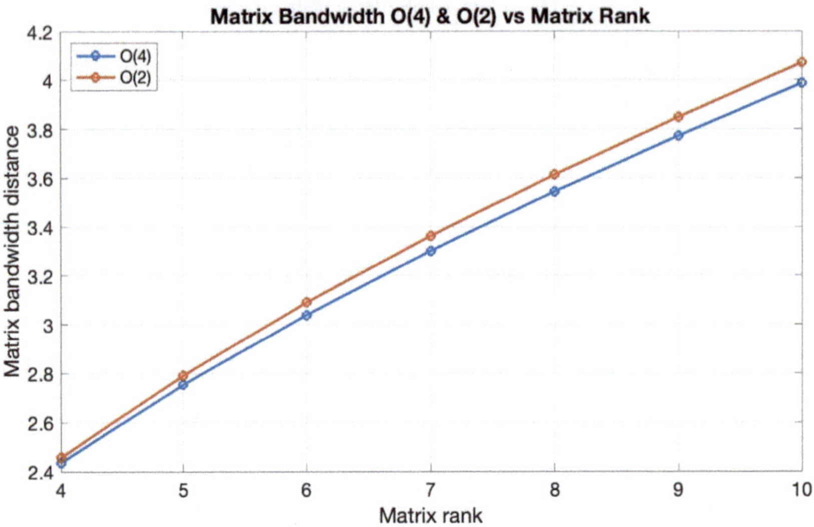

Fig. 9.3 Matrix bandwidth distance versus matrix rank for different fractional powers of the operator matrix and approximation orders. $O(4)$ versus $O(2)$ approximation for p/q power= 4/5

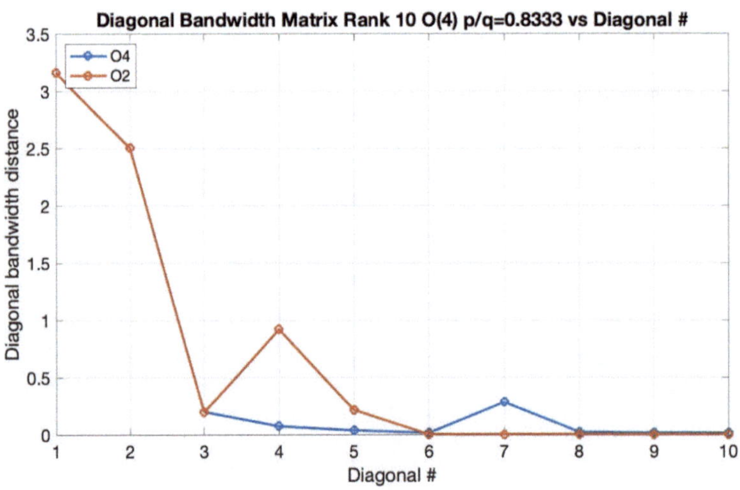

Fig. 9.4 Diagonal bandwith distance versus matrix diagonal # for two approximation orders $O(4)$ and $O(2)$ and p/q power = 5/6

On one hand, we can evaluate the fractional derivative of order $\frac{1}{2}$ of the exponential function, $\mathbf{x}(t) = \mathbf{e}^t$, at 7 discrete points (every 1 unit from 3 to 9), this is $\frac{\mathbf{d}^{0.5}}{\mathbf{d}t^{0.5}}\mathbf{x}(t)$, (with a sum of up to 50 terms) obtaining the following vector,

$$\frac{\mathbf{d}^{0.5}}{\mathbf{d}t^{0.5}}\mathbf{x}(t) = \begin{pmatrix} 7.452 \\ 20.124 \\ 54.625 \\ 148.433 \\ 403.443 \\ 1096.646 \\ 2980.969 \end{pmatrix} \tag{9.39}$$

On the other hand, the corresponding fractional differential matrix operator of order $\frac{1}{2}$ for a rank of N = 7 and approximation error $O(4)$ from Eq. (9.36) was,

$$\mathbf{D}^{0.5} = \frac{1}{\Delta t^{0.5}} \begin{pmatrix} 1 & 0 & 0 & 0 & 0 & 0 & 0 \\ -0.5000 & 1 & 0 & 0 & 0 & 0 & 0 \\ -0.1250 & -0.5000 & 1 & 0 & 0 & 0 & 0 \\ -0.0625 & -0.1250 & -0.5000 & 1 & 0 & 0 & 0 \\ -0.0391 & -0.0625 & -0.1250 & -0.5000 & 1 & 0 & 0 \\ -0.0271 & -0.0391 & -0.0625 & -0.1250 & -0.5000 & 1 & 0 \\ -0.0635 & -0.0271 & -0.0391 & -0.0625 & -0.1250 & -0.5000 & 1 \end{pmatrix} + O(4) \tag{9.40}$$

The bandwidth distance of this matrix operator in this case corresponds to d = 2.9332 and when we compute the operation over $\mathbf{x}$(t) for Δt = 1.0, the following vector is obtained,

$$\mathbf{D}^{0.5}\mathbf{x}(t) = \begin{pmatrix} 6.03 \\ 16.051 \\ 43.462 \\ 118.035 \\ 320.779 \\ 872.139 \\ 2370.756 \end{pmatrix}_{\Delta t=1} \tag{9.41}$$

We can observe that the latter result fulfills the similitude between the two differential operators, $\mathbf{D}^{0.5}\mathbf{x}(t) \approx \frac{\mathbf{d}^{0.5}}{\mathbf{d}t^{0.5}}\mathbf{x}(t)$, when $\mathbf{x}(t) = \mathbf{e}^t$. However, it lacks precision near a 20.3% root mean square error. However, when $\Delta t = 0.795$, the RMS error reduces considerably down to 0.3% and the resulting vector elements are closer to the ones of the former solution.

$$\mathbf{D}^{0.5}\mathbf{x}(t) = \begin{pmatrix} 7.585 \\ 20.19 \\ 54.669 \\ 148.472 \\ 403.496 \\ 1097.030 \\ 2982.083 \end{pmatrix}_{\Delta t=0.795} \tag{9.42}$$

9.9 Conclusions

- Measurements on a system must precede its mathematical modeling. Data from experiments will usually be available in the form of a matrix-vector, from which we will have to come up with a mathematical model of the system. For this reason, it is useful to work with matrices instead of functions.
- Here we evaluate a first-order derivative operator **D** square matrix raised to a fractional power. We identify a suitably-defined bandwidth distance of the matrix, and observe that it increases with the rank of the matrix and with the order of the derivative, towards unity; being always below the bandwidth distance of the first-order derivative operator associated matrix.
- We also evaluated the bandwidth distance of each of the none-zero diagonals of the operator **D** square matrix raised to a fractional power. As the diagonal is further apart from the main diagonal, the bandwidth distance values decrease almost exponentially and it is sensitive to the approximation $O()$ used to compute the matrix.
- In this particular study, the $sqrtm()$, $expm(\ )$ and $logm(\ )$ functions are part of the MATLAB library of advanced mathematical functions, and they compare to the $O(4)$ approximation for the $exp()$ and $ln()$ infinite series functions presented here. The $O(2)$ approximation shows a slightly higher deviation from the latter approximation regarding the bandwidth distances.
- It is interesting to observe that for a backward-in-time first-order differential matrix operator, when raised to a fractional power, the lower triangle of the matrix gets completely populated indicating a possible memory effect of this specific fractional matrix operator.
- The backward-in-time differential operator when raised to a fractional power between 0 and 1, provides a closed approximation (but not precise) to the actual fractional derivative of the same order range when applied to the exponential function when the latter is evaluated from a power series.
- Next, we will work on the relation between the elevation of a matrix to a fractional power and the corresponding derivative of fractional order as well as specific applications regarding physical experiments and their associated data.

Competing Interests The authors have no conflicts of interest to declare that are relevant to the content of this chapter.

Acknowledgements This study was funded by the Academic Vicerectory and the School of Engineering of PUC as well as by the Aerospace and Mechanical Engineering Department of UND.

References

1. Hedayati M, Ezzati R (2013) A new operational matrix method to solve nonlinear fractional differential equations. Nonlinear Eng Model Appl 13(1)
2. Kazem S, Abbasbandy S, Kumar S (2013) Fractional-order Legendre functions for solving fractional-order differential equations. Appl Math Model 37(17):5498–5510
3. Tarasov VE (2011) Fractional dynamics: applications of fractional calculus to dynamics of particles. Springer, Fields and Media
4. Fallahgoul H, Focardi S, Fabozzi F (2016) Fractional calculus and fractional processes with applications to financial economics: theory and application. Academic Press
5. Evangelista LR (2023) Fractional diffusion equations and anomalous diffusion. Cambridge University Press, Cambridge, U.K
6. Higham N (2008) Functions of matrices: theory and computation. SIAM, Philadelphia

Competing Interests The authors have no conflicts of interest to declare that are relevant to the content of this chapter.

Acknowledgements [illegible]

References

[illegible]

Chapter 10
Analysis of the Fractional Schrödinger Equation with Free Particle Potential in One Dimension with Caputo and Caputo-Fabrizio Derivatives

Pedro Oliva-Sanchez, Ruben Aguilar-Marquez, Javier Alejandro Pérez-Garza, Edelmiro Leal-Fernandez, and Servando Lopez-Aguayo

10.1 Introduction

Fractional calculus is a branch of mathematics that generalizes derivatives and integrals to non-integer orders. It originated in 1695 when L'Hôpital sent a letter to Leibniz seeking the meaning of a half-order derivative. However, unlike integer or traditional derivatives, the fractional derivative can be defined in various ways because it is a non-local operator. As a result, numerous definitions of the fractional derivative with different characteristics have been proposed in recent years, including the Riemann-Liouville, Caputo, Grünwald-Letnikov, and conformable definitions, among many others with distinct features.

These tools have proven extremely useful in modeling systems with memory properties, such as in the modeling of viscoelastic materials [1], anomalous diffusion [2], fluid mechanics [3], among others [4–6]. In general, fractional derivatives allow us to describe the dynamics of systems with atypical behavior more accurately. In quantum mechanics, for example, the Schrödinger equation plays a fundamental role in describing quantum systems. However, under certain conditions, such as systems with anomalous diffusion [7] or complex media, modeling with the Schrödinger equation may not provide a solution that truly approximates the system's dynamics. In this sense, introducing a fractional version of the Schrödinger equation can more accurately capture the solution framework for modeling the system.

Therefore, it is interesting to explore the fractional Schrödinger equation in one dimension more deeply and understand the behavior of quantum systems that obey

P. Oliva-Sanchez · R. Aguilar-Marquez · J. A. Pérez-Garza · S. Lopez-Aguayo (✉)
Tecnológico de Monterrey, Escuela de Ingeniería y Ciencias, Monterrey, N.L., Mexico
e-mail: servando@tec.mx

E. Leal-Fernandez
Universidad Autonoma de Nuevo León, Facultad de Ciencias Físico, Pedro de Alba, Niños Héroes, Ciudad Universitaria, San Nicolás de los Garza, N.L., Mexico

J. J. Rosales García et al. (eds.), *Advances in Fractional Calculus*,
Modeling and Optimization in Science and Technologies 21,
https://doi.org/10.1007/978-3-031-83989-4_10

this formulation. The objective of this work is to present an approach to the fractional Schrödinger equation in one dimension. In particular, the physical implications of introducing different definitions of fractional spatial derivatives will be analyzed. The analytical solutions obtained will be expressed in terms of special functions that depend on the fractional order, providing a more detailed view of the effects these modifications have on the quantum dynamics of particles in simple potentials.

10.2 Some Definitions of Fractional Calculus

In recent years, various definitions of fractional derivatives and integrals have been proposed, where the value of α represents the order of the fractional derivative.

The Caputo derivative is one of the most relevant and widely accepted definitions in the scientific community, defined as [8]:

$$ {}_a^C D_x^\alpha f(x) = \frac{1}{\Gamma(n-\alpha)} \int_a^x \frac{f^{(n)}(\tau)}{(x-\tau)^{\alpha+1-n}} d\tau, \tag{10.1} $$

where $n-1 < \alpha < n$, that is, $n = \lceil \alpha \rceil$, and $\Gamma(\alpha)$ is the Gamma function. The Laplace transform of the Caputo derivative is given by:

$$ \mathfrak{L}\left\{ {}_a^C D_x^\alpha f(x) \right\} = s^\alpha F(s) - \sum_{k=0}^{n-1} s^{\alpha-k-1} f^{(k)}(0), \tag{10.2} $$

For this definition, the derivative of a constant is zero, which allows us to define the initial conditions of fractional differential equations analogously to the classical integer-order case.

Another recent definition is the Caputo-Fabrizio fractional derivative:

$$ {}^{CF} D_x^\alpha f(x) = \frac{M(\alpha)}{1-\alpha} \int_a^t f'(\tau) e^{-\frac{\alpha(t-\tau)}{1-\alpha}} d\tau, \tag{10.3} $$

where $M(\alpha)$ is a normalization function such that $M(0) = M(1) = 1$ and $\alpha \in [0, 1]$ [9]. The Laplace transform for this definition of fractional derivative is given by:

$$ \mathfrak{L}\left\{ {}^{CF} D_x^\alpha f(x) \right\} = \frac{sF(s) - f(0)}{s + \alpha(1-s)}, \tag{10.4} $$

Given the above definitions, it is important to clarify that a basic criterion must be met to be considered a fractional derivative. In this paper, we consider the criteria proposed by Ortigueira and Machado [10]: (i) Linearity: The fractional derivative operator must satisfy the linearity criterion. (ii) Identity: The fractional derivative of order zero must return the original function. (iii) Compatibility with integers: A fractional derivative must coincide with the original derivative when the order is an integer. (iv) Order preservation: It must be satisfied that $D^\alpha D^\beta f(t) = D^{\alpha+\beta} f(t)$, for $\alpha, \beta > 0$. Finally, (v) The generalized Leibniz rule:

$$D^\alpha[f(t)g(t)] = \sum_{j=0}^{\infty} \binom{\alpha}{j} D^j f(t) D^{\alpha-j} g(t).$$

which generalizes the product rule.

However, there is some controversy in the literature regarding the definition used by Caputo and Fabrizio, as Ortigueira and Machado point out in [11] that this definition does not satisfy criterion (iv) and does not behave as a fractional operator under the Laplace transform. Despite this, in this work, we use the Caputo-Fabrizio definition due to its particular characteristics: convergence, non-singular kernel, and behavior that depends on the order of the derivative. Although we acknowledge that it does not meet all the formal criteria for being a fractional derivative, we consider that its properties are beneficial for the specific analysis of the fractional Schrödinger equation in this study.

10.3 Fractional Schrödinger Equation

The Schrödinger equation is one of the fundamental pillars of quantum mechanics; this equation describes how the quantum state of a system evolves over time. In general, the Schrödinger equation has the form:

$$i\hbar \frac{\partial \Psi}{\partial t} = -\frac{\hbar^2}{2m} \nabla^2 \Psi + V\Psi, \tag{10.5}$$

where Ψ is the wave function that describes the quantum system, V is the potential acting on the particle, and $\hbar$ and m are the reduced Planck constant and the mass of the particle, respectively. This equation contains all the information about the system to be studied. In the case of one dimension, the Schrödinger equation takes a more manageable form:

$$i\hbar \frac{\partial \Psi}{\partial t} = -\frac{\hbar^2}{2m} \frac{\partial^2 \Psi}{\partial x^2} + V\Psi, \tag{10.6}$$

With this simplification, it has been possible to describe systems such as harmonic oscillators, potential wells, and potential barriers, among other cases.

To fractionalize the order of a differential equation, different methods exist. Typically, the transition from the ordinary differential equation to the fractional differential equation is done directly. Another approach to fractionalizing a differential equation is presented in [12], where a parameter σ is proposed to avoid affecting the units of the differential equation. However, to avoid the determination of this parameter, in this article, the variables have been nondimensionalized, and the ordinary derivative has been directly replaced with a fractional derivative. Since the variables are nondimensionalized, this approach prevents issues with the units in the solution. The nondimensionalized Schrödinger equation has the form:

$$i\frac{\partial \Psi}{\partial \tilde{t}} = -\frac{1}{2}\frac{\partial^2 \Psi}{\partial \tilde{x}^2} + V\Psi, \tag{10.7}$$

where $\tilde{t}$ and $\tilde{x}$ are the nondimensionalized coordinates, now we can write this fractional differential equation in the following way:

$$i\frac{\partial \Psi}{\partial \tilde{t}} = -\frac{1}{2}\frac{\partial^{2\alpha} \Psi}{\partial \tilde{x}^{2\alpha}} + V\Psi. \tag{10.8}$$

This fractionalization will be used with the Caputo definition, as it is commonly applied in this form. The form that the differential equation can take using the Caputo-Fabrizio definition can be expressed as:

$$i\frac{\partial \Psi}{\partial \tilde{t}} = -\frac{1}{2}\frac{\partial^{1+\alpha} \Psi}{\partial \tilde{x}^{1+\alpha}} + V\Psi. \tag{10.9}$$

For the Caputo-Fabrizio definition, an ordinary derivative and a fractional derivative are usually used to express higher orders. Considering Ψ as a separable function of the form $\Psi(\tilde{x}, \tilde{t}) = \psi(\tilde{x})\phi(\tilde{t})$, the solution to Eqs. (10.8) and (10.9) is $\phi(\tilde{t}) = e^{-iE\tilde{t}}$, so taking a potential for free particle $V = 0$, they can be rewritten as:

$$\frac{d^{2\alpha}\psi}{d\tilde{x}^{2\alpha}} = -k^2\psi. \tag{10.10}$$

$$\frac{d^{1+\alpha}\psi}{d\tilde{x}^{1+\alpha}} = -k^2\psi. \tag{10.11}$$

where $k = \sqrt{2E}$.

For Eq. (10.10), the solutions, taking the initial conditions as $\psi(0) = \psi_0$ and $\psi'(0) = 0$, are given by:

$$\psi(\tilde{x}) = \psi_0 \mathbf{E}_{2\alpha}\left\{-k^2\tilde{x}^{2\alpha}\right\}, \tag{10.12}$$

where

$$\mathbf{E}_{2\alpha}\left\{-k^2\tilde{x}^{2\alpha}\right\} = \sum_{n=0}^{\infty}\frac{(-k^2\tilde{x}^{2\alpha})^n}{\Gamma(2\alpha n + 1)},$$

stands for the Mittag-Leffler function, so for Eq. (10.8), the solutions take the form:

$$\Psi(\tilde{x},\tilde{t}) = \psi_0 \mathbf{E}_{2\alpha}\left\{-k^2\tilde{x}^{2\alpha}\right\} e^{-\frac{ik^2}{2}\tilde{t}}, \tag{10.13}$$

In Fig. 10.1, different forms that Eq. (10.13) can take are shown. For the case $\alpha = 1$, the Mittag-Leffler function converges to a hyperbolic cosine of the form $\cosh(\sqrt{z})$, so the equation takes the form:

$$\Psi(\tilde{x},\tilde{t}) = \psi_0 \cosh(\sqrt{-k^2 E\tilde{x}^2})e^{-\frac{ik^2}{2}\tilde{t}},$$

$$= \psi_0 \cos(k\tilde{x})e^{-\frac{ik^2}{2}\tilde{t}}.$$

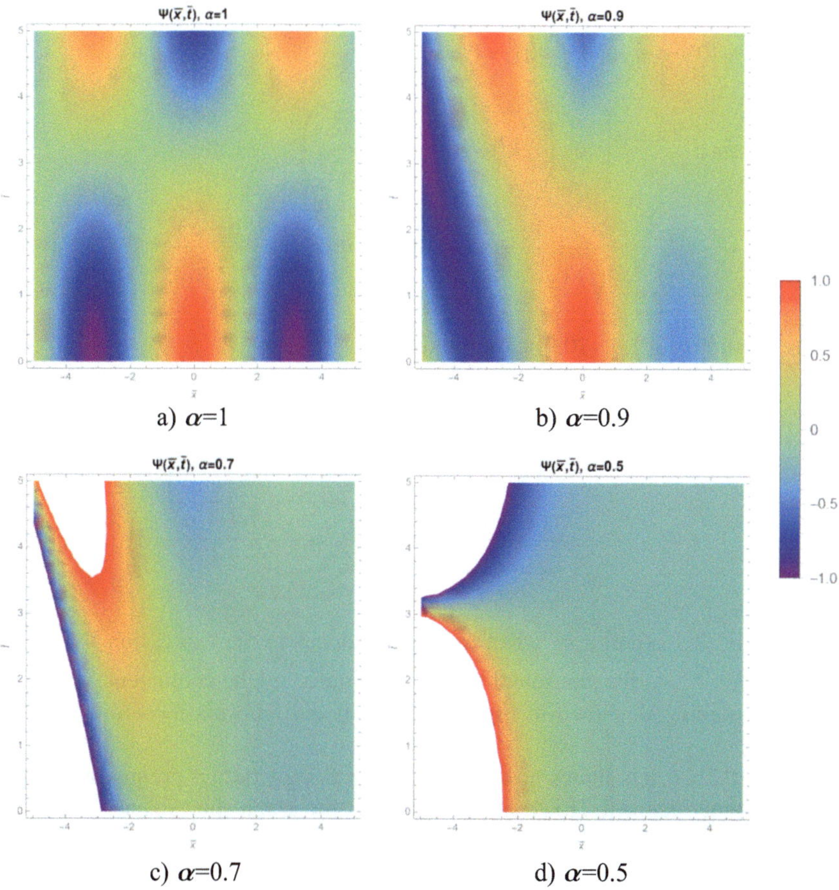

Fig. 10.1 Solutions for the Schrödinger equation with Caputo derivative for $\alpha = 1, \alpha = 0.9, \alpha = 0.7, \alpha = 0.5$

Therefore, in the case $\alpha = 1$, the ordinary solution for the wave function is recovered, as shown in Fig. 10.1a. This figure represents the expected behavior without visible perturbations, where the wave function retains its symmetry and periodicity.

Figure 10.1b shows an inclined asymmetry, which might indicate a trend in displacement. It also illustrates how oscillations reduce in amplitude, which could be related to damping or energy loss in the system. In Fig. 10.1c, the function displays regions that grow and decay indefinitely; this behavior may be linked to effects potentially associated with the fractional nature of the derivative used in the equation, similar to Fig. 10.1d. These changes suggest that variations in the order introduce perturbations that alter boundary conditions or system properties. They imply that variations in α affect the dynamic characteristics of the system.

On the other hand, we have Eq. (10.11), which corresponds to the expression for the Caputo-Fabrizio derivative. The solutions for this equation with the initial conditions $\psi(0) = \psi_0$ and $\psi'(0) = 0$ are:

$$\psi(\tilde{x}) = \psi_0 e^{-\frac{k^2(1-\alpha)}{2}\tilde{x}} \left[\cos\left(\sqrt{k^2\alpha - \frac{k^4(1-\alpha)^2}{4}}\tilde{x} \right) - \frac{k^2(1-\alpha)}{2\sqrt{k^2\alpha - \frac{k^4(1-\alpha)^2}{4}}} \sin\left(\sqrt{k^2\alpha - \frac{k^4(1-\alpha)^2}{4}}\tilde{x} \right) \right], \quad (10.14)$$

The solutions for Eq. (10.9) have the form:

$$\Psi(\tilde{x},\tilde{t}) = \psi_0 e^{-\frac{ik^2}{2}\tilde{t}} e^{-\frac{k^2(1-\alpha)}{2}\tilde{x}} \left[\cos\left(\sqrt{k^2\alpha - \frac{k^4(1-\alpha)^2}{4}}\tilde{x} \right) - \frac{k^2(1-\alpha)}{2\sqrt{k^2\alpha - \frac{k^4(1-\alpha)^2}{4}}} \sin\left(\sqrt{k^2\alpha - \frac{k^4(1-\alpha)^2}{4}}\tilde{x} \right) \right]. \quad (10.15)$$

Similarly, in this expression, when $\alpha = 1$, it reduces to the expression $\Psi(\tilde{x},\tilde{t}) = \psi_0 \cos(k\tilde{x}) e^{-\frac{ik^2}{2}\tilde{t}}$, as the sinusoidal term is eliminated by the coefficient dependent on α it contains. As shown in Fig. 10.2a, the model recovers the ordinary wave function.

In Fig. 10.2b, c, it is shown how the oscillation along $\tilde{x}$ decays progressively until it is completely damped. However, in Fig. 10.2c, the graph shows zones where the function grows and decays rapidly. This behavior may be associated with effects within the system. In contrast, Fig. 10.2d shows that for the value $\alpha = 0$. Similarly to the Caputo derivative, the graph exhibits regions that grow and decay indefinitely. These behaviors suggest a transition between simpler and more complex dynamics as α varies.

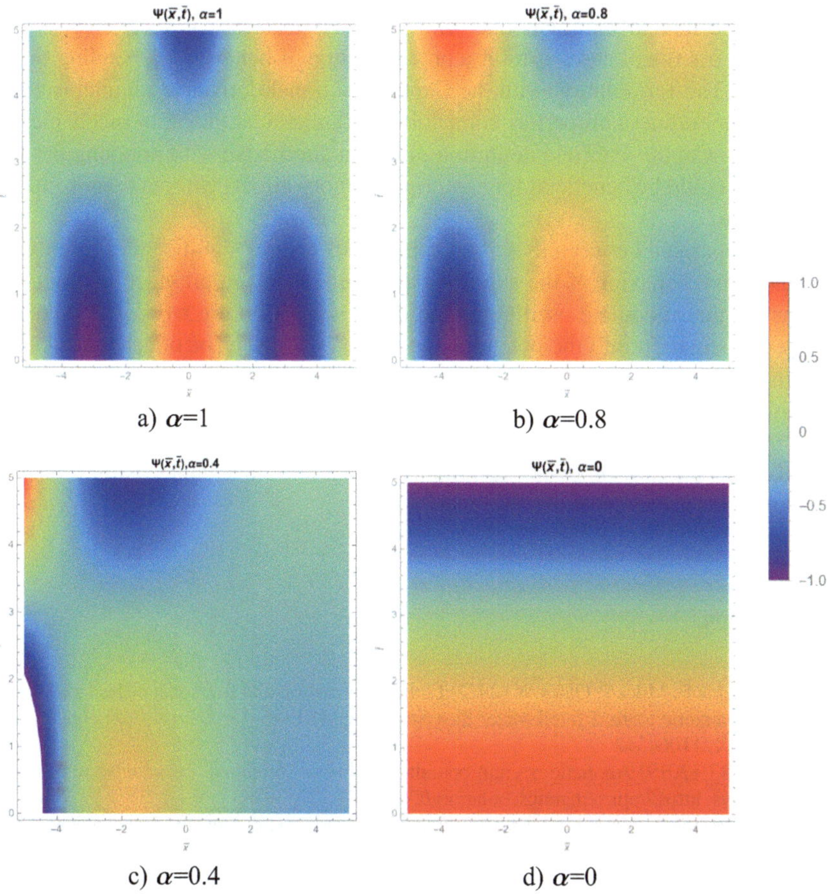

Fig. 10.2 Solutions for the Schrödinger equation with Caputo-Fabrizio derivative for $\alpha = 0$, $\alpha = 0.4$, $\alpha = 0.8$, $\alpha = 1$

10.4 Conclusion

In this work, an analytical exploration of the fractional Schrödinger equation in one dimension has been presented, employing different definitions of fractional spatial derivatives. Specifically, we have analyzed the implications of adopting the Caputo and Caputo-Fabrizio derivative definitions. By directly transforming the standard Schrödinger equation into its fractional counterparts, we derived new Eqs. (10.10) and (10.11), which characterize the wave function in simple potentials under different fractional orders, α. The solutions, expressed in terms of special functions such as the Mittag-Leffler function, provide a detailed view of how fractional modifications affect quantum dynamics.

It is important to highlight that while the classical case is recovered by setting $\alpha = 1$, fractional cases exhibit a broader range of behaviors depending on the value of α. The formulation with the Caputo fractional derivative maintains continuity and smoothness properties compatible with classical quantum mechanics. In contrast, the Caputo-Fabrizio definition offers an alternative with non-singular kernels, which could be beneficial for capturing memory effects without compromising convergence.

The comparison between these two approaches reveals significant differences in the qualitative behavior of the solutions, particularly in the presence of fractional parameters that modulate oscillatory and decay terms. Future work will involve a deeper investigation into the potential applications of these fractional formulations in complex quantum systems and an analysis of their stability and convergence properties under various boundary conditions. This study aims to contribute to a better understanding of the underlying mechanics of fractional quantum systems and pave the way for new modeling techniques in quantum mechanics using fractional calculus.

References

1. Stanisauskis E, Miles P, Oates W (2022) Finite deformation and fractional order viscoelasticity of an auxetic foam. J Intell Mater Syst Struct 33(14):1846–1861. https://doi.org/10.1177/1045389X211064342
2. Sibatov RT (2019) Anomalous grain boundary diffusion: fractional calculus approach. Adv Math Phys. https://api.semanticscholar.org/CorpusID:127638583
3. Lazopoulos KA, Lazopoulos AK (2017) Fractional vector calculus and fluid mechanics. J Mech Behav Mater 26:43–54. https://api.semanticscholar.org/CorpusID:126221728
4. Tarasov VE (2011) Fractional dynamics: applications of fractional calculus to dynamics of particles, fields and media. Retrieved from https://api.semanticscholar.org/CorpusID:118534045
5. Hilfer R (2000) Applications of fractional calculus in physics. Retrieved from https://api.semanticscholar.org/CorpusID:117406670
6. Sun H, Zhang Y, Baleanu D, Chen W, Chen YQ (2018) A new collection of real world applications of fractional calculus in science and engineering. Commun Nonlinear Sci Numer Simul 64:213–231. Retrieved from https://api.semanticscholar.org/CorpusID:125743874
7. Chen W (2006) Time–space fabric underlying anomalous diffusion. Chaos, Solitons Fract 28(4):923–929. Sir Hermann Bondi 1919–2005. https://doi.org/10.1016/j.chaos.2005.08.199
8. Caputo M (1969) Elasticità e Dissipazione. Zanichelli, Bologna
9. Caputo M, Fabrizio M (2015) A new definition of fractional derivative without singular kernel. Progress Fract Differ Appl 1(2). https://doi.org/10.1016/j.jcp.2014.07.019
10. Ortigueira MD, Tenreiro Machado JA (2015) What is a fractional derivative? J Comput Phys 293:4–13. Fractional PDEs. https://doi.org/10.1016/j.jcp.2014.07.019
11. Ortigueira MD, Tenreiro Machado J (2018) A critical analysis of the Caputo–Fabrizio operator. Commun Nonlinear Sci Numer Simul 59:608–611. https://doi.org/10.1016/j.cnsns.2017.12.001
12. Gómez-Aguilar JF, Rosales-García JJ, Bernal-Alvarado JJ, Córdova-Fraga T, Guzmán-Cabrera R (2012) Fractional mechanical oscillators. Revista mexicana de física 58(4):348–352

Chapter 11
Fractionalization of Hermite-Gaussian Optical Beams by a Fourier Approach

Ruben Aguilar-Marquez, Javier Alejandro Pérez-Garza, Arath Maldonado-Traconis, Pedro Oliva-Sanchez, and Servando Lopez-Aguayo

11.1 Introduction

Hermite-Gaussian beams naturally solve the paraxial equation in Cartesian coordinates and provide a complete orthonormal basis for complex optical fields. They are used in applications such as parabolic index multimode fibers [1], super-resolution imaging [2], and optical trapping [3]. While the relationships between different integer orders of Hermite-Gaussian beams are established through derivatives, the role of fractional calculus is less explored [4].

There are several definitions of fractional derivatives, each with specific mathematical properties, as detailed in the next section. These definitions can lead to varying results, allowing researchers to select the one that best matches their experimental data.

Beyond being mathematical curiosities, fractional derivatives play a vital role in scientific and engineering problems. Typically, we use classical derivatives to describe evolution or propagation equations due to the concept of instantaneous rate of change, as seen in Markovian processes. However, to study processes with memory or non-instantaneous effects-like anomalous diffusion or evolution in nonlocal media-fractional derivatives provide a natural mathematical model. In recent years, fractional calculus has become valuable in various areas of physics, including electrodynamics [5], quantum mechanics [6, 7], quantum field theory [8], to mention a few. In addition to this, fractional calculus has found its way into engineering and applied mathematics, for purposes such as RC circuit analysis [9, 10], machine learning [11–13], computer vision [14–16] and control theory [17].

Fractional calculus has become influential in optics. For example, the fractional Fourier transform is used to describe light distribution in optical systems with

R. Aguilar-Marquez · J. A. Pérez-Garza · A. Maldonado-Traconis · P. Oliva-Sanchez · S. Lopez-Aguayo (✉)
Tecnológico de Monterrey, Escuela de Ingeniería y Ciencias, Monterrey, N.L., Mexico
e-mail: servando@tec.mx

J. J. Rosales García et al. (eds.), *Advances in Fractional Calculus*, Modeling and Optimization in Science and Technologies 21,
https://doi.org/10.1007/978-3-031-83989-4_11

lenses [18]. Gutierrez-Vega applied fractional derivatives to the paraxial wave equation [19], while fractional calculus has proven useful in modelling soliton propagation in nonlocal media [20, 21]. It has also been applied to nonlinear Schrödinger equations to model Lévy flights in optical cavities [22]. These applications highlight fractional calculus's effectiveness in modeling nonlocal phenomena and memory effects [23, 24].

In this work, we explore optical beams as a bridge between classical integer-order beams, focusing on the fractionalization of generalized Hermite-Gaussian beams. We use the Fourier fractional derivative, which aligns well with optical contexts. The paper is structured as follows: Sect. 11.2 covers basic fractional derivative definitions, Sect. 11.3 shows that the fractional Fourier derivative meets Ortigueira-Machado criteria, Sect. 11.4 introduces fractional Hermite polynomials, Sect. 11.5 presents generalized fractional Hermite-Gaussian (GFHG) beams and their conditions for existence, Sect. 11.6 discusses GFHG beams in physical systems, Sect. 11.7 shows some properties of the GFHG beams, Sect. 11.8 presents numerical simulations of GFHG beam propagation, and Sect. 11.9 concludes the study.

11.2 Definition and Properties of Fractional Derivatives

This section presents alternative definitions of fractional derivatives beyond the fractional Fourier derivative. The parameter α denotes the fractional order, which need not be an integer.

One common definition is the Riemann-Liouville fractional derivative [25]:

$$ {}_{a}^{RL}D_x^{\alpha} f(x) = \frac{1}{\Gamma(m-\alpha)} \left(\frac{d}{dx}\right)^{m} \int_a^x (x-\tau)^{m-\alpha-1} f(\tau) d\tau, \tag{11.1} $$

where $m = \text{floor}(\alpha)$ and $m \leq \alpha < m+1$.

Another definition is the Caputo fractional derivative [26]:

$$ {}_{a}^{C}D_x^{\alpha} f(x) = \frac{1}{\Gamma(n-\alpha)} \int_a^x \frac{f^{(n)}(\tau)}{(x-\tau)^{\alpha+1-n}} d\tau, \tag{11.2} $$

where $n = \text{ceil}(\alpha)$ and $n-1 < \alpha < n$.

The Caputo-Fabrizio fractional derivative is defined as [27]:

$$ {}^{CF}D_x^{\alpha} f(x) = \frac{M(\alpha)}{1-\alpha} \int_a^x f'(\tau) e^{-\frac{\alpha(x-\tau)}{1-\alpha}} d\tau, \tag{11.3} $$

with $M(\alpha)$ being a normalization function and $\alpha \in [0, 1]$.

Other definitions include the Atangana-Baleanu Caputo fractional derivative [28]:

$$^{ABC}\mathrm{D}^{\alpha}_{a+} f(x) = \frac{M(\alpha)}{1-\alpha} \int a^{x} \mathrm{E}\alpha \left(-\frac{\alpha(x-\tau)^{\alpha}}{1-\alpha} \right) f'(\tau) d\tau,$$

the conformable derivative [28]:

$$\mathrm{D}^{\alpha} f(x) = \lim_{\epsilon \to 0} \frac{f(x+\epsilon x^{1-\alpha}) - f(x)}{\epsilon},$$

and the Grünwald-Letnikov derivative:

$${}^{GL}_{a}D^{\alpha} f(x) = \lim_{h \to 0} h^{-\alpha} \sum_{k=0}^{\frac{x-a}{h}} (-1)^{k} \binom{\alpha}{k} f(x-kh).$$

To evaluate fractional derivatives, we consider the criteria proposed by Ortigueira and Machado [29]:

- Linearity: The operator must be linear.
- Identity: The zero-order derivative must return the original function.
- Backward Compatibility: It should match the classical derivative for integer orders.
- Law of Orders: $D^{\alpha} D^{\beta} f(t) = D^{\alpha+\beta} f(t)$ for $\alpha, \beta > 0$.
- Generalized Leibniz Rule:

$$D^{\alpha}[f(t)g(t)] = \sum_{j=0}^{\infty} \binom{\alpha}{j} D^{j} f(t) D^{\alpha-j} g(t), \tag{11.4}$$

which generalizes the product rule.

However, some definitions with non-singular kernels, like the Caputo-Fabrizio derivative, face controversy. Ortigueira and Machado [30] argue that this definition does not satisfy the last criterion and behaves differently from other fractional operators in its Laplace transform.

11.3 Ortigueira-Machado Criteria for the Fractional Fourier Derivative

As indicated in Sect. 11.2, the criteria for determining the validity of the definition of fractional derivatives are those proposed by Ortigueira and Machado [29]. This section demonstrates how the fractional Fourier derivative given by Eq. (11.7) satisfies these criteria.

- Linearity: Since this fractional derivative is directly dependent on the Fourier transform and the inverse Fourier transform, both of which are linear, this property is immediately observable.
- Index law: By applying the definition provided in Eq. (11.7), the result is:

$$^{F}\mathrm{D}_x^{\alpha}[^{F}\mathrm{D}_x^{\beta}[f(x)]] = \mathcal{F}^{-1}\left[(ik)^{\beta}(ik)^{\alpha}\hat{f}(k)\right],$$

Similarly,

$$^{F}\mathrm{D}_x^{\beta}[^{F}\mathrm{D}_x^{\alpha}[f(x)]] = \mathcal{F}^{-1}\left[(ik)^{\alpha}(ik)^{\beta}\hat{f}(k)\right],$$

also,

$$^{F}\mathrm{D}_x^{\alpha+\beta}[f(x)] = \mathcal{F}^{-1}\left[(ik)^{\alpha+\beta}\hat{f}(k)\right],$$

therefore, we can deduce that:

$$^{F}\mathrm{D}_x^{\alpha}[^{F}\mathrm{D}_x^{\beta}[f(x)]] =^{F} \mathrm{D}_x^{\beta}[^{F}\mathrm{D}_x^{\alpha}[f(x)]] =^{F} \mathrm{D}_x^{\alpha+\beta}[f(x)].$$

- Identity: Considering $\beta = -\alpha$ we get

$$^{F}\mathrm{D}_x^{\alpha}[^{F}\mathrm{D}_x^{-\alpha}[f(x)]] = \mathcal{F}^{-1}\left[(ik)^{0}\hat{f}(k)\right] = f(x).$$

- Backward compatibility: By definition, the fractional Fourier derivative when n is an integer coincides with the standard derivative.
- The generalized Leibniz rule: This derivative satisfies the generalized Leibniz rule as shown in [31] where the expression is:

$$D^{\alpha}[f(t)g(t)] = \frac{i^{\alpha}}{2\pi}\int_{-\infty}^{\infty} e^{isx}\hat{g}(s)\left(\int_{-\infty}^{\infty} e^{iqx}\hat{f}(q)(s+q)^{\alpha}dq\right)ds,$$

The above expression can be rewritten by expanding the binomial as a Taylor series in the form$\sum_{k=0}^{\infty}\binom{\alpha}{k}q^k s^{\alpha-k}$ and finally, we obtain

$$D^{\alpha}[f(t)g(t)] = \frac{i^{\alpha}}{2\pi}\sum_{k=0}^{\infty}\binom{\alpha}{k}\int_{-\infty}^{\infty} e^{isx}\hat{g}(s)s^{\alpha-k}\left(\int_{-\infty}^{\infty} e^{iqx}\hat{f}(q)q^k dq\right)ds,$$

by rearranging the coefficient $\frac{i^{\alpha}}{2\pi}$ as $\left(\frac{i^k}{\sqrt{2\pi}}\right)\left(\frac{i^{\alpha-k}}{\sqrt{2\pi}}\right)$. We recover the definition of Fourier fractional derivative $D_t^{\alpha-k}g(t) = \mathcal{F}^{-1}\{(is)^{\alpha-k}\hat{g}(s)\}$ and $D_t^k f(t) = \mathcal{F}^{-1}\{(iq)^k\hat{f}(q)\}$. Thus we get the generalized Leibniz rule.

$$D^{\alpha}[f(t)g(t)] = \sum_{k=0}^{\infty}\binom{\alpha}{k}D^k f(t)D^{\alpha-k}g(t).$$

therefore the condition is fulfilled.

Since the fractional Fourier derivative meets the criteria proposed in [29], under these conditions we can argue that the Fourier definition corresponds to a fractional derivative operator.

11.4 Fractional Hermite Functions

To begin, we generalize the Hermite functions. The Hermite polynomials $H_\alpha(x)$ are defined using the Rodrigues formula [32] $H_\alpha(x) = (-1)^\alpha e^{x^2} \frac{d^\alpha}{dx^\alpha} e^{-x^2}$. Following the method proposed by Fourier to compute a derivative of arbitrary order using Fourier transforms [4], let us define the Fourier transform as

$$\mathcal{F}\{f(x)\} = \hat{f}(k) = \frac{1}{\sqrt{2\pi}} \int_{-\infty}^{\infty} f(x) e^{-ikx} \mathrm{d}x, \tag{11.5}$$

and the inverse Fourier transform as

$$\mathcal{F}^{-1}\{\hat{f}(k)\} = f(x) = \frac{1}{\sqrt{2\pi}} \int_{-\infty}^{\infty} \hat{f}(k) e^{ikx} \mathrm{d}k, \tag{11.6}$$

Then, the derivative can be defined as

$$\frac{d^\alpha}{dx^\alpha} f(x) = \frac{1}{\sqrt{2\pi}} \int_{-\infty}^{\infty} (ik)^\alpha \hat{f}(k) e^{ikx} \mathrm{d}k, \tag{11.7}$$

We focus on the case where the parameter α can be any positive real value. The next step is to use fractional calculus to determine the fractional derivative of the Hermite functions $\frac{d^\alpha}{dx^\alpha} e^{-\frac{x^2}{B^2}}$, denoted as $D_x^\alpha e^{-\frac{x^2}{B^2}}$, where B is constant. By defining the Fourier transform as in Eq. (11.5), the property shown in Eq. (11.7) can be exploited to find $D_x^\alpha f(x)$ by using the inverse Fourier transform as $D_x^\alpha f(x) = \mathcal{F}^{-1}\{(ik)^\alpha \mathcal{F}\{f(x)\}\}$. Thus, we can define:

$$(-1)^\alpha D_x^\alpha e^{-\frac{x^2}{B^2}} = \mathcal{F}^{-1}\left\{(-ik)^\alpha \mathcal{F}\left\{e^{-\frac{x^2}{B^2}}\right\}\right\}. \tag{11.8}$$

Calculating that $\mathcal{F}\left\{e^{-\frac{x^2}{B^2}}\right\} = \frac{B}{\sqrt{2}} e^{-\frac{1}{4}B^2k^2}$, the integral form of this equation is described by

$$(-1)^\alpha D_x^\alpha e^{-\frac{x^2}{B^2}} = \frac{B}{2\sqrt{\pi}} \int_{-\infty}^{\infty} (-ik)^\alpha \, e^{-\frac{1}{4}B^2k^2 + ikx} \, dk. \tag{11.9}$$

By working out Eq. (11.9), and substituting into the Rodrigues Formula, we obtain

$$H_\alpha(x) = A\frac{2^\alpha}{\sqrt{\pi}}e^{\frac{x^2}{B^2}}\vartheta_\alpha\left(\frac{x}{B}\right). \tag{11.10}$$

where A is a constant, and $\vartheta_\alpha(x)$ is given by

$$\begin{aligned}\vartheta_\alpha(x) = \Gamma\left(\frac{1+\alpha}{2}\right){}_1F_1\left(\frac{1+\alpha}{2},\frac{1}{2},-x^2\right)\cos\left(\frac{\pi\alpha}{2}\right)+\\ x\alpha\Gamma\left(\frac{\alpha}{2}\right){}_1F_1\left(1+\frac{\alpha}{2},\frac{3}{2},-x^2\right)\sin\left(\frac{\pi\alpha}{2}\right).\end{aligned} \tag{11.11}$$

here ${}_1F_1(a,b,z)$ is the Kummer's (confluent hypergeometric) function. By using the property that [32],

$${}_1F_1(a,b,z) = e^z\,{}_1F_1(b-a,b,-z), \tag{11.12}$$

then $\vartheta_\alpha\left(\frac{x}{B}\right)$ is equivalent to $e^{-\frac{x^2}{B^2}}I_\alpha\left(\frac{x}{B}\right)$, where $I_\alpha\left(\frac{x}{B}\right)$ is given by

$$\begin{aligned}I_\alpha(x) = \Gamma\left(\frac{1+\alpha}{2}\right){}_1F_1\left(-\frac{\alpha}{2},\frac{1}{2},x^2\right)\cos\left(\frac{\pi\alpha}{2}\right)+\\ x\alpha\Gamma\left(\frac{\alpha}{2}\right){}_1F_1\left(\frac{1-\alpha}{2},\frac{3}{2},x^2\right)\sin\left(\frac{\pi\alpha}{2}\right).\end{aligned} \tag{11.13}$$

With this result, the fractionalization of the Hermite functions using the Fourier definition results in

$$H_\alpha(x) = A\frac{2^\alpha}{\sqrt{\pi}}I_\alpha\left(\frac{x}{B}\right). \tag{11.14}$$

The goal of introducing parameters A and B into this fractional function is to facilitate their use as adjustment parameters in optimization algorithms, which often need multiple evaluations. Thus, the expression should be easy and quick to evaluate. However, we found confluent hypergeometric functions inefficient for this. Instead, applying a fractional integral representation of the Rodrigues formula [32] yields an equivalent integral form, replacing the classical integer parameter with a fractional one. This gives us

$$H_\alpha(x) = A\frac{2^{\alpha+1}}{\sqrt{\pi}}\exp(x^2)\int_0^\infty \exp(-t^2)\cos(2xt-\frac{\pi}{2}\alpha)\,dt. \tag{11.15}$$

This expression, equivalent to Eq. (11.14) is a key result of this manuscript. Numerical experiments show it is not only easier to program but also two orders of magnitude faster to compute than the analytical expression with confluent hypergeometric functions. Figure 11.1 illustrates fractional Hermite functions: Fig. 11.1a shows classical

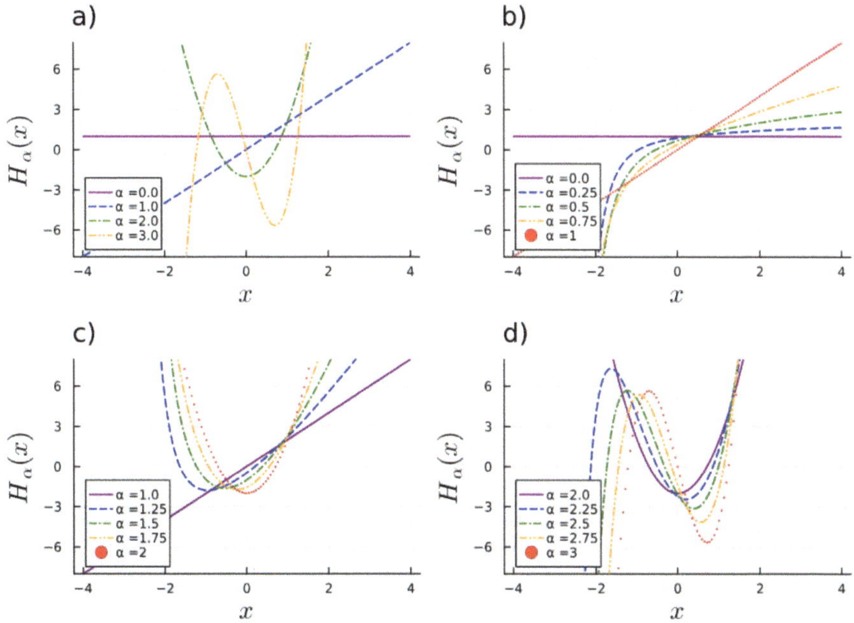

Fig. 11.1 Fractional Hermite functions. **a** Integer Hermite functions, followed by fractional Hermite functions for the cases where α is between **b** 0 and 1, **c** 1 and 2, and, **d** 2 and 3

integer-order Hermite functions, while Fig. 11.1b displays fractional Hermite functions from 0 to 1, highlighting increased asymmetry at negative x values. Figure 11.1c, d depict transitions between $H_1(x)$ and $H_2(x)$, and $H_2(x)$ and $H_3(x)$, respectively. The asymmetry in how the functions approach $\pm\infty$ is evident and can be analytically verified through the exponential term in Eq. 11.12.

11.5 Generalized Hermite-Gaussian Beams

Next, we apply apodization to ensure the localization of our initial fields. The Gaussian function, commonly used in the paraxial wave equation, serves as the classical apodization. Therefore, we introduce the factor $\exp(-x^2/C^2)$, where C is a constant and, finally, we get the generalized fractional 1D Hermite Gaussian (GFHG) beams described by

$$HG_\alpha(x) = A\frac{2^\alpha}{\sqrt{\pi}}I_\alpha(x/B)\exp(-x^2/C^2); \tag{11.16}$$

For simplicity, we omit the explicit dependence of A, B, and C in the function, i.e., $HG_\alpha(x) = HG_\alpha(x; A, B, C)$. Here, α denotes the fractional order, bridging traditional Hermite-Gaussian functions. Parameter A controls the maximum amplitude,

B scales the transverse dimension, and C defines the Gaussian apodization width. Unlike integer Hermite-Gaussian beams, we identified a threshold value between B and C that is crucial for generating GFHG beams. This is illustrated by the limiting form of the confluent hypergeometric function, ${}_1F_1\,(a, b, z)$ as $x \to \infty$ [33].

$$ {}_1F_1\,(a, b, x) \sim \frac{1}{\Gamma\,(a)} e^x x^{a-b}, \tag{11.17}$$

except in polynomial cases when $a = 0, -1, -2, \ldots$.

By writing Eq. (11.14) in its limiting form as $x \to \infty$, Eq. (11.16) can be written

$$ HG_\alpha(x) \sim A\frac{2^{\alpha-1}}{\sqrt{\pi^3}} e^{\frac{x^2}{B^2}-\frac{x^2}{C^2}} \sin\,(\pi\alpha)\,\Gamma\left(\frac{\alpha}{2}\right)\Gamma\left(\frac{1+\alpha}{2}\right)\left(\frac{x}{B}\right)^{-\frac{\alpha}{2}-\frac{1}{2}}\left[\left(\frac{x}{B}\right)^{\frac{1}{2}} - \frac{1}{2}\right]. \tag{11.18}$$

Next, we analyze the following scenarios:

- $B > C$: In this case, the expression asymptotically approaches $\exp[-(B^2 - C^2)/C^2B^2]$ as $|x| \to \infty$, causing $HG_\alpha(x)$ to converge to 0 as $|x| \to \infty$, resulting in a bounded function.
- $B = C$: Conversely, in this case, the exponential term has no contribution giving as a result that $HG_\alpha(x)$ tends to 0 as $|x| \to \infty$ for fractional n.
- $B < C$: For this last case, the expression is asymptotic to $\exp[+(C^2 - B^2)/C^2B^2]$ as $|x| \to \infty$, causing $HG_\alpha(x)$ to diverge due the exponential term grows faster than the rest of the function that converges to 0.

Thus, a necessary condition for the existence of fractional Hermite-Gaussian beams, as described in Eq. (11.16), is that $B \geq C$. This condition applies only to fractional α values and real B and C. This is the second main result of this manuscript, demonstrating that the two rescaling parameters for the Hermite and Gaussian functions cannot be freely chosen in the fractional case. We present the profiles of the GFHG beams, with Fig. 11.2a showing classical integer Hermite-Gaussian beams. The transitions between integer orders are illustrated for HG_0 to HG_1, HG_1 to HG_2, and HG_2 to HG_3 in Fig. 11.2b–d, respectively. Note that GFHG beams lack both odd and even symmetry, unlike their integer counterparts.

11.6 Fractional Hermite-Gaussian Beams as Solution to Physical Systems

Until now, we have focused on a mathematical framework. However, a natural question arises: are these fractional functions valid for describing physical optical fields? Indeed, these fractional functions can describe solutions to the scalar paraxial wave equation, which can be expressed as

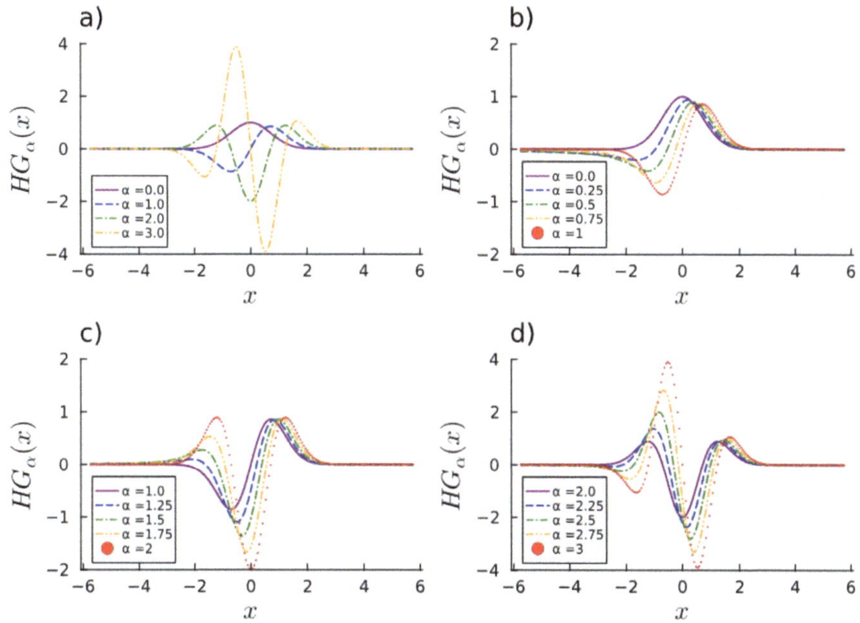

Fig. 11.2 Fractional Hermite-Gaussian beams: **a** Integer orders, while examples of transitions are shown for **b** HG_0 and HG_1, **c** HG_1 and HG_2 and **d** HG_2 and HG_3. In all figures, $B = 1$ and $C = 1$. A condition for the existence of these beams, ensuring localization, is that $B \geq C$

$$i2K\frac{\partial U(x,z)}{\partial z} + \frac{\partial U^2(x,z)}{\partial x^2} + V(x,z)U(x,z) = 0, \tag{11.19}$$

here K stands for the wave vector, z and x represent the longitudinal and transverse spatial coordinates, respectively, $U(x, z)$ is the complex optical field, and $V(x, z)$ is the potential that describes the properties of the propagation medium. For instance, $V(x, z) = 0$ corresponds to free space. This equation has a fundamental solution in free space described by

$$U(x,z) = \frac{A_1}{q(z)}\exp\left(-\frac{iKx^2}{2q(z)}\right), \tag{11.20}$$

where $q(z)$ is the q-parameter of the beam, given by $q(z) = z + iz_0$, with z_0 being the Rayleigh length, and A_1 is a normalization constant. By Schwarz's theorem, if U_o is a solution to the paraxial wave equation, then $\partial^n U_o/\partial x^n$ is also a solution. Thus, using a linear fractional operator, we can apply generalized fractional Hermite-Gaussian functions to describe optical beam propagation in a fractional form. For free space propagation, we can set $B = C = \omega_0\sqrt{\mu}$, where $\mu \equiv \mu(z) = 1 + i\zeta$, and $\zeta \equiv \zeta(z) = \frac{z}{z_0}$, with $z_0 = \frac{K\omega_0^2}{2}$ as the Rayleigh length. In contrast, for other media, such as graded index (GRIN) media, $B \neq C$, requiring specific relationships for parameters A, B, and C. These can be determined using variational methods, as with Hermite-Gaussian beams in complex systems like nonlocal nonlinear media [34].

Siegman introduced elegant Hermite-Gaussian beams (eHGB) as an alternative to standard Hermite-Gaussian beams (sHGB), noting that sHGBs lack symmetry between the complex argument of the Gaussian function and the real argument of the Hermite function [35].

With $B = C = \omega_0\sqrt{\mu}$, we can construct an exact solution to the paraxial equation in free space, known as eHGB, which remain well-bounded as $|x| \rightarrow \infty$ for fractional α.

$$U_e(x,z) = HG_\alpha\left(x, z; \frac{A_\alpha\sqrt{\omega_0^\alpha}e^{i(1+\alpha)\Psi(z)}}{\sqrt{w(z)^{\alpha+1}}}, \omega_0\sqrt{\mu}, \omega_0\sqrt{\mu}\right). \tag{11.21}$$

where the amplitude is in terms of a normalization constant A_α, the radius of the beam $w(z) = \omega_0(1+\zeta^2)^{\frac{1}{2}}$, and the Gouy phase shift $\Psi(z) = -\frac{1}{2}\arctan(\zeta)$.

On the other hand, we can also build an exact solution to the paraxial equation for free space with the sHGB, which diverge for fractional n,

$$U_s(x,z) = HG_n\left(x, z; \frac{A_\alpha e^{i(1+2\alpha)\Psi(z)}}{\sqrt{w(z)}}, \frac{w(z)}{\sqrt{2}}, \omega_0\sqrt{\mu}\right). \tag{11.22}$$

In this case, we need to analyze Eq. (11.18) for $B = \frac{w(z)}{\sqrt{2}}$, and $C = \omega_0\sqrt{\mu}$. By taking $C^2 = \omega_0 w(z)e^{i\arctan(\zeta)}$, and working $e^{\frac{x^2}{B^2}-\frac{x^2}{C^2}}$ we get that the n for this case must always be an integer number, as the fractional n will cause divergence in both the real, and the imaginary parts of the function.

Siegman notes that the width of eHGB at its waist is reduced, significantly altering its diffraction and phase structure compared to sHGB. As a result, eHGBs lose their structural stability during propagation and do not maintain the usual Gaussian spherical-phase variation, meaning they are no longer shape-invariant, unlike standard HGB [35]. This may explain why eHGB can propagate in a fractionalized form, while sHGB cannot. The differences between eHGBs and sHGBs under fractionalization raise several questions for future research: How stable are these optical beams with different B and C scaling parameters? If the fractionalization procedure changes, do these answers remain valid?

Moreover, Hermite-Gaussian functions have applications beyond optics; the findings here can be relevant in other fields. For example, consider a quantum mechanics model describing a particle subject to a restoring force proportional to its displacement, akin to a mass on a spring in classical mechanics-the quantum harmonic oscillator described by

$$-\frac{\hbar^2}{2m}\frac{\partial^2\psi(x)}{\partial x^2} + [V(x) - E_n]\,\psi(x) = 0. \tag{11.23}$$

where in this particular case the potential $V(x) = \frac{1}{2}m\omega^2 x^2$ and an eigenenergy of $E_n = \hbar\omega\left(n + \frac{1}{2}\right)$. Here n is a non-negative integer, $\hbar$ is the reduced Planck's constant, m is the mass of the particle, ω is the angular frequency of the oscillator, and the wave function $\psi(x)$ that solves this equation is indeed, in form of the Hermite-Gaussian functions described by

$$\psi_n(x) = HG_n\left(x; A_n\left(\frac{m\omega}{\hbar}\right)^{\frac{1}{4}}, \sqrt{\frac{\hbar}{m\omega}}, \sqrt{\frac{2\hbar}{m\omega}}\right). \tag{11.24}$$

Using the result reported earlier in this paper, since $C > B$, we conclude that the parameter n must be an integer. This also provides an alternative demonstration that the quantum harmonic oscillator cannot be fractionalized and only allows discrete values. Therefore, GFHG beams can be used to study not only well-known optical systems but also new, more complex ones, and other types of physical systems.

11.7 Properties of the Generalized Fractional Hermite-Gaussian Beams

Theoretically, classical higher-order Hermite-Gaussian optical beams can be generated using creation and annihilation operators. Alternatively, using the fractional Fourier derivative, fractional Hermite-Gaussian functions can be generated through a nonlocal or spatial memory-like process by convolving a fundamental Gaussian beam with the appropriate kernel $M(x)$, or equivalently

$$HG_\alpha(x) = \left[\int_{-\infty}^{\infty} M(x-\tau)HG_0(\tau)d\tau\right]\exp(-x^2/D^2). \tag{11.25}$$

where $1/D^2 = (B^2 - C^2)/(C^2B^2)$, $HG_0(x) = \exp(-x^2/B^2)$, and $M(x)$ is given by $M(x) = \mathcal{F}^{-1}\{(-ik)^\alpha\}$, given as a result:

$$M(x) = \begin{cases} (-1)^\alpha\sqrt{2\pi}\delta^{(\alpha)}(x), & \text{if } \alpha \in \mathbb{N}. \\ \frac{\sin(\pi\alpha)\Gamma(\alpha+1)}{\sqrt{2\pi}}\frac{(\text{sgn}(x)-1)}{|x|^{\alpha+1}}, & \text{otherwise.} \end{cases} \tag{11.26}$$

Here, $\delta^{(\alpha)}(x)$ represents the α-derivative of the Dirac delta function, and $sgn(x)$ is the sign function. The expressions in Eqs. (11.25) and (11.26) represent the third key result of this manuscript. They show that higher-order fractional GFHG beams can be understood as interactions between the fundamental mode and a nonlocal response characterized by a power law. Interestingly, the Fourier transform of k^α in Eq. (11.7) also follows a power law, $1/|x|^{\alpha+1}$, as similar to the fact that the Fourier transform of a Hermite Gauss function is another Hermite-Gaussian function. Thus, the beams introduced here may help establish connections between optical beams,

Fourier optics, and nonlocal media. Nonlinear nonlocal media, such as photorefractive or thermal optical media, can be represented through a convolution between the medium and the optical beam propagating through it, similar to the convolution expression shown in Eq. (11.25).

Another notable property is the discovery of recurrence relations in the fractional case, allowing the generation of higher-order fractional beams using:

$$HG_{\alpha+1}(x) = \frac{2x}{B} HG_{\alpha}(x) - 2\alpha HG_{\alpha-1}(x), \tag{11.27}$$

this relation holds for any α, as long as B and C are equal in the three terms. Another relation involving derivatives is

$$HG'_{\alpha}(x) + \frac{2x}{C^2} HG_{\alpha}(x) = \frac{2\alpha}{B} HG_{\alpha-1}(x), \tag{11.28}$$

again, this relation holds for any α, as long as B and C are equal in the three terms.

11.8 Propagation of the Generalized Fractional Hermite-Gauss Beams

Finally, we conduct numerical propagations to illustrate some properties of the beams that can be analyzed using Eq. (11.16). We model optical propagation in free space described by Eq. (11.19), setting $V(x, z) = 0$ and employing a numerical pseudo-spectral propagation method. For comparison purposes, in Fig. 11.3a, we show the evolution of a simple Gaussian beam, or equivalently, a fractional Hermite-Gaussian beam with $\alpha = 0$, $B = 1$, $C = 1/\sqrt{3}$. In contrast , we set $B = C = 1$ and propagate a GFHG beam of $\alpha = 1.5$ for a distance of $z = 0.5$ as shown in Fig. 11.3c. The beam naturally undergoes diffraction, but due to the nature of the Hermite functions, certain sections remain invariant over a longer propagation distance, especially the central peak. This behavior, similar to that of Bessel-Gauss beams, shows that GFHG beams are more robust to diffraction, akin to integer Hermite-Gaussian beams [36]. Thus, future studies could explore the self-healing and non-diffracting behavior of the GFHG beams. Next, we compare our numerical propagations with the analytical expression of the GFHG beams in free space, as given by Eq. (11.21). The comparison between the analytical expression and the numerical result is illustrated for $z = 0.25$ in HG_0 in Fig. 11.3b and for $HG_{1.5}$ in Fig. 11.3d, respectively. Observe that in both cases, the analytical beam generated by Eq. 11.17 and the beam obtained from direct numerical propagation are equivalent, respectively. Similar behavior is observed for all values of z, demonstrating the usefulness of the analytical expression.

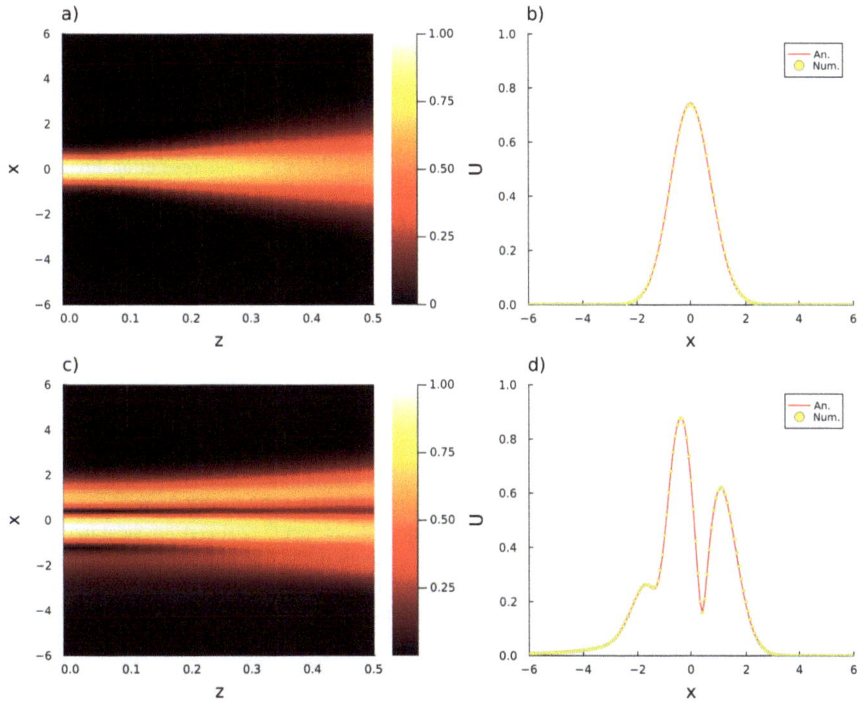

Fig. 11.3 Propagation of generalized fractional Hermite-Gaussian beams. We show propagation in free space for HG_0 with $B = 1$, $C = 1/\sqrt{3}$ in **a**, and for $HG_{1.5}$ with $B = C = 1$ in **c**. Respectively, we show the comparison between Eq. (11.21) at $z = 0.25$ and the numerical propagation

11.9 Conclusions

In conclusion, in this work, we introduced generalized fractional Hermite-Gaussian beams, which offer a flexible approach for modeling optical beams under the paraxial approximation. These beams are defined by four parameters: α, the fractional order connecting traditional Hermite-Gaussian beams; A, the normalization parameter, important in nonlinear propagation; B, which scales the transverse dimension; and C, which scales Gaussian apodization. Localization of these fields requires $B \geq C$, a condition not necessary for integer orders. Unlike integer beams, GFHG beams lack odd or even symmetry but retain several mathematical relationships, including those between fractional orders. We demonstrate their use in studying free-space propagation and suggest their application to more complex media. We hope these findings will inspire further exploration of fractional optical beams.

Acknowledgements The authors thank Julio Cesar Gutierrez-Vega for useful comments.

References

1. Zhou Y, Braverman B, Fyffe A, Zhang R, Zhao J, Willner AE, Shi Z, Boyd RW (2021) High-fidelity spatial mode transmission through a 1-km-long multimode fiber via vectorial time reversal. Nat Commun 12:1866
2. Frank J, Duplinskiy A, Bearne K, Lvovsky A (2023) Passive super resolution imaging of incoherent objects. Optica 10:1147
3. Shukri M, Thabit F (2024) Calculation of trapping optical forces induced by a focused continuous Hermite Gaussian beam on a nano-dielectric spherical particle. Physica Scripta 99:065504
4. Herrmann R (2018) Fractional calculus: an introduction for physicists, 3rd edn. World Scientific
5. Tarasov VE (2022) General non-local electrodynamics: equations and non-local effects. Ann Phys 445. https://doi.org/10.1016/j.aop.2022.169082
6. Laskin N (2000) Fractional quantum mechanics and Lévy path integrals. Phys Lett A 268. https://doi.org/10.1016/S0375-9601(00)00201-2
7. Longhi S (2015) Fractional Schrödinger equation in optics. Opt Lett 40. https://doi.org/10.1364/ol.40.001117
8. Atman KG, Şirin H (2020) Nonlocal phenomena in quantum mechanics with fractional calculus. Rep Math Phys 86. https://doi.org/10.1016/S0034-4877(20)30075-6
9. Gómez-Aguilar JF, Razo-Hernández R, Granados-Lieberman D (2014) A physical interpretation of fractional calculus in observables terms: analysis of the fractional time constant and the transitory response. Revista Mexicana de Fisica 60 (2014)
10. Rosales JJ, Filoteo JD, González A (2018) A comparative analysis of the RC circuit with local and non-local fractional derivatives. Revista Mexicana de Fisica 64. https://doi.org/10.31349/REVMEXFIS.64.647
11. Raubitzek S, Mallinger K, Neubauer T (2023) Combining fractional derivatives and machine learning: a review. https://doi.org/10.3390/e25010035
12. Zhang H, Pu Y-F, Xie X, Zhang B, Wang J, Huang T (2021) A global neural network learning machine: coupled integer and fractional calculus operator with an adaptive learning scheme. Neural Netw 143:386
13. Joshi M, Bhosale S, Vyawahare VA (2023) A survey of fractional calculus applications in artificial neural networks. Artif Intell Rev 56:13897
14. Arora S, Mathur T, Agarwal S, Tiwari K, Gupta P (2022) Applications of fractional calculus in computer vision: a survey. Neurocomputing 489:407
15. Marques F, De Araujo TP, Nator C, Saraiva A, Sousa J, Pinto AM, Melo R (2019) Recognition of simple handwritten polynomials using segmentation with fractional calculus and convolutional neural networks. In: 2019 8th Brazilian conference on intelligent systems (BRACIS). IEEE, pp 245–250
16. Yazgaç BG, Kırcı M (2019) Fractional order calculus based fruit detection. In: 2019 8th international conference on agro-geoinformatics (Agro-Geoinformatics). IEEE, pp 1–4
17. Aburakhis M, Ordóñez R (2024) Generalization of direct adaptive control using fractional calculus applied to nonlinear systems. J Control Autom Electr Syst
18. Ozaktas H, Mendlovic D (1995) Fractional fourier optics. JOSA A 12:743–751
19. Gutiérrez-Vega JC (2007) Fractionalization of optical beams: I planar analysis. Opt Lett 32:1521. https://doi.org/10.1364/ol.32.001521
20. Zeng H, Wang Y, Xiao M, Wang Y (2023) Fractional solitons: new phenomena and exact solutions. Front Phys 11. https://doi.org/10.3389/fphy.2023.1177335
21. Briedis D, Petersen D, Edmundson D, Krolikowski W, Bang O (2005) Ring vortex solitons in nonlocal nonlinear media. Opt Express 13:435. https://doi.org/10.1364/OPEX.13.000435
22. Malomed B (2021) Optical solitons and vortices in fractional media: a mini-review of recent results. Photonics 8:353
23. Maghsoudi-Khouzani S, Kurt A (2024) New semi-analytical solution of fractional Newell–Whitehead–Segel equation arising in nonlinear optics with non-singular and non-local kernel derivative. Opt Quant Electron 56. https://doi.org/10.1007/s11082-023-06126-4

24. Holm S (2023) Natural occurrence of fractional derivatives in physics. In: 2023 International conference on fractional differentiation and its applications (ICFDA), pp 1–5. https://doi.org/10.1109/ICFDA58234.2023.10153355
25. Guía-Calderón M, Rosales-García JJ, Guzmán-Cabrera R, González-Parada A, Álvarez Jaime JA (2015) El cálculo diferencial e integral fraccionario y sus aplicaciones. Acta Universitaria 25:20. https://doi.org/10.15174/au.2015.688
26. Caputo M (1969) Elasticitàe Dissipazione. Zanichelli, Bologna
27. Caputo M, Fabrizio M (2015) A new definition of fractional derivative without singular kernel Prog Fract Differ Appl 1. https://doi.org/10.12785/pfda/010201
28. Valério D, Ortigueira MD, Lopes AM (2022) How many fractional derivatives are there? Mathematics 10:737
29. Ortigueira MD, Tenreiro Machado JA (2015) What is a fractional derivative? J Comput Phys 293:4–13. https://doi.org/10.1016/j.jcp.2014.07.019
30. Ortigueira MD, Tenreiro Machado J (2018) A critical analysis of the Caputo-Fabrizio operator. Commun Nonlinear Sci Numer Simul 59:608. https://doi.org/10.1016/j.cnsns.2017.12.001
31. Bagarello F (2019) Fourier transforms, fractional derivatives, and a little bit of quantum mechanics. ArXiv Preprint arXiv:1912.01836. Available via https://arxiv.org/pdf/1912.01836
32. Abramowitz M, Stegun IA (1964) Handbook of mathematical functions with formulas, graphs, and mathematical tables. Dover Publications
33. DLMF (2024) Confluent hypergeometric functions: Kummer functions. National Institute of Standards and Technology. Available via https://dlmf.nist.gov/13.2E23
34. Buccoliero D, Desyatnikov AS, Krolikowski W, Kivshar YS (2007) Laguerre and Hermite soliton clusters in nonlocal nonlinear media. Phys Rev Lett 98:053901
35. Siegman A (1973) Hermite-Gaussian functions of complex argument as optical-beam eigenfunctions. J Opt Soc Am 63(9):1093–1094
36. Bencheikh A, Forbes A (2020) The non-diffracting nature of truncated Hermite-Gaussian beams. JOSA A 37:C1

26. Holm S (2023) Natural occurrence of fractional derivatives in physics. In: 2023 International conference on fractional differentiation and its applications (ICFDA) pp 1–6. https://doi.org/10.1109/ICFDA58234.2023.10153355

27. [illegible]

28. [illegible]

29. [illegible]

Chapter 12
Fractional Models of the Catenary Curve

Leonardo Martínez-Jiménez, Jorge Mario Cruz-Duarte, J. Juan Rosales-García, and Miroslava Cano-Lara

12.1 Introduction

The application of the catenary curve on construction of arches is attributed to Robert Hooke. Although the catenary is a concept originated in mechanic, it has been applied in civil structures, railway electrical systems [1], human dental arch [2, 3], recently, it has been shown that it has many more applications in optics, such as large solar collectors, geodesic antennas and membrane lenses [4] . In [5] investigated the effect of catenary action on the progressive collapse potential of steel moment framed structures.

On the other hand, recently, the FC has had a strong impulse through the development of new definitions and applications. FC is a generalization of the ordinary calculus, when the derivatives and integrals have non-integer orders. Nowadays, it has been used in several studies due to its features to model many processes in different areas of science and engineering [6–9]. Fractional derivatives are generally defined by integrals and therefore are non-local operators, so they have memory.

L. Martínez-Jiménez (✉)
Departamento de Estudios Multidisciplinarios, División de Ingenierías, Campus Irapuato-Salamanca, Universidad de Guanajuato, Yuriria, Gto., Mexico
e-mail: leonardomj@ugto.mx

J. M. Cruz-Duarte
Escuela de Ingeniería y Ciencias, Tecnológico de Monterrey, Monterrey, NL, Mexico
e-mail: jorge.cruz@tec.mx

J. J. Rosales-García
Departamento de Ingeniería Eléctrica, División de Ingenierías Campus Irapuato–Salamanca, Universidad de Guanajuato, Salamanca, Gto., Mexico
e-mail: rosales@ugto.mx

M. Cano-Lara
Departamento de Mecatrónica, TecNM/ITS Irapuato, Irapuato, Gto., Mexico
e-mail: miroslava.cl@irapuato.tecnm.mx

J. J. Rosales García et al. (eds.), *Advances in Fractional Calculus*,
Modeling and Optimization in Science and Technologies 21,
https://doi.org/10.1007/978-3-031-83989-4_12

Nonlocal distribution effects are essential for the accurate description and understanding of some complex dynamical systems. There are many classic applications where FC has proven its capabilities. For example, the fractional diffusion equation [10], linear capacitor theory [11], non-local description of quantum dynamics [12], fault monitoring systems for induction motors [13], medical imaging: [14–16], analysis of mechanical models [17], models of synchronization [18], models on cosmology [19], caotic systems [20], control of mobil robots [21], analysis of electrical circuits [22]. However, due to the absence of a unique physical and geometric interpretation of the fractional derivative, various definitions have been proposed and implemented [23]. Some of them are Riemann–Liouville (RL), Grünwald–Letnikov (GL), CD, Weyl (W) [24–27], the Caputo-Fabrizio Fractional Derivative (CF) [28], the Atangana–Balenau Fractional Derivative (ABC) [29] and the local non-integer Fractional Conformable Derivative [30]. Some works have proposed the study of the properties of the catenary curve when using fractional operators: in 2019 [31], a fractional catenary model was proposed using the CF operator and, in 2024, [32], described a fractional model of the catenary was implemented using the Caputo fractional derivative and compare the solution with real physical models.

The aim of this work is to give an analysis of the fractional catenary curve using some fractional definitions. Particularly, we will use the best known fractional derivatives, such as: CD and ABC, as well as the conformable fractional derivative. The paper is organized as follows: Sect. 12.2 presents the definitions and some properties of the operators that are used; Sect. 12.3 reviews the catenary curve's model; Sect. 12.4 shows the proposed CD and ABC fractional models, as well as the Conformable model of the catenary curve; Sect. 12.5 shows the obtained results. Finally, Sect. 12.6 discusses the results, and some conclusions are commented.

12.2 Basic Concepts

12.2.1 The CD Fractional Derivative

The CD fractional derivative of a continuous function, $f(x) : \mathbb{R} \to \mathbb{R}$ is defined as follows [33],

$$\frac{d^\alpha f}{dx^\alpha} = {}^{\mathrm{CD}}D_x^\alpha f = \frac{1}{\Gamma(1-\alpha)} \int_a^x (x-\tau)^{-\alpha} \left(\frac{df}{dx}\right)_{x=\tau} d\tau, \tag{12.1}$$

with $\alpha \in (0, 1]$, $a \in (-\infty, x]$, and $f \in H^1(a, b)$, $b > a$. Its corresponding Laplace transform is given by

$$\mathcal{L}\left[{}^{\mathrm{CD}}D_x^\alpha f(x)\right] = s^\alpha F(s) - s^{\alpha-1} f(0). \tag{12.2}$$

12.2.2 The ABC Derivative

The ABC fractional derivative is defined as follows [29],

$$ {}^{ABC}D_x^{\alpha} f(x) = \frac{B(\alpha)}{1-\alpha} \int_a^x f'(\tau) E_\alpha \left(-\alpha \frac{(x-\tau)^\alpha}{1-\alpha} \right) d\tau. \tag{12.3} $$

where the kernel $(x-\tau)^{-\alpha}$ in (12.1) is replaced by $E_\alpha\left(-\alpha\frac{(x-\tau)^\alpha}{1-\alpha}\right)$. Unlike the Caputo-Dzhrbashyan fractional derivative (12.1), the ABC derivative has no singularity at $x=\tau$. Its Laplace transform is determined by

$$ \mathcal{L}[{}^{ABC}D_x^{\alpha} f(x)] = \frac{B(\alpha)}{1-\alpha} \cdot \frac{s^\alpha F(s) - s^{\alpha-1} f(0)}{s^\alpha + \frac{\alpha}{1-\alpha}}, \tag{12.4} $$

where $B(\alpha)$ is a normalization function, usually taken equal to unity. $E_\alpha(\cdot)$ in (12.3) is the one-parameter ML function given by [34–36]

$$ E_\alpha(x) = \sum_{k=0}^{\infty} \frac{x^k}{\Gamma[\alpha k + 1]}, \quad \alpha \in \mathbb{C}, \quad Re(\alpha) > 0, \tag{12.5} $$

and the generalized ML function of three parameters is

$$ E_{\alpha,\beta}^{\delta}(x) = \sum_{k=0}^{\infty} \frac{(\delta)_k}{\Gamma[\alpha k + \beta]} \frac{x^k}{k!}, \quad \alpha, \beta \in \mathbb{C}, \quad Re(\alpha), \; Re(\beta) > 0, \tag{12.6} $$

where $(\delta)_k$ is the Pochhammer symbol. It is noteworthy that $E_{\alpha,\beta}^{1}(x) = E_{\alpha,\beta}(x)$ and $E_{1,1}(x) = e^x$, so the ML functions are called generalizations of the exponential function.

In terms of Laplace transform, the ML function is found such as,

$$ \mathcal{L}[x^{\beta-1} E_{\alpha,\beta}^{\delta}(-a x^\alpha)] = \frac{s^{\alpha\delta-\beta}}{(s^\alpha + a)^\delta}, \tag{12.7} $$

$\alpha, \beta > 0$, and $a \in \mathbb{R}$, (12.7) constitutes a powerful relationship between space (and/or time) and complex domains. In addition, a useful property of the ML function is

$$ E_{-\alpha,\beta}(x) = -x^{-1} E_{\alpha,\alpha+\beta}\left(x^{-1}\right). \tag{12.8} $$

12.2.3 Conformable Derivative

The Conformable derivative, defined as a local and non-integer derivative in [30], is presented as follows,

$$T_\alpha f(x) = f^\alpha(x) = \lim_{\epsilon \to 0} \frac{f(x + \epsilon x^{1-\alpha}) - f(x)}{\epsilon}, \tag{12.9}$$

for all $x > 0$, where $f : [0, \infty) \to \mathbb{R}$ is a function and $\alpha \in (0, 1]$ is the non-integer order. When $\alpha = 1$, the conformable derivative in (12.9) becomes $f'(x) = \lim_{\epsilon \to 0} \frac{f(x+\epsilon)-f(x)}{\epsilon}$, as expected. Likewise, if f is α-differentiable in some $(0, a)$, $a > 0$, and $\lim_{x \to 0^+} f^\alpha(x)$ exists, then it is defined

$$f^\alpha(0) = \lim_{x \to 0^+} f^\alpha(x). \tag{12.10}$$

Finally, the most important properties of this derivative are listed in [30]. We will use only the property

$$T_\alpha f(x) = x^{1-\alpha} \frac{df(x)}{dx}, \quad \alpha \in (0, 1], \tag{12.11}$$

if $f(x)$ is differentiable at $x > 0$.

12.3 Catenary Curve

The catenary curve is the shape described by a cable or string hanging naturally due to its own weight, and it is supported at its extremes. The catenary curve has some important features that described it, *i.e.*, the minimum height y_0 over floor is at the point $p_1 = (x_0, y_0)$, and an arbitrary point p_2 is located at (x, y), since x and y are the horizontal and vertical axes, respectively. The cable weight in the segment $\overline{p_1 p_2}$ is assumed as $\lambda L g$, where λ is the linear density, L is the arc length and g is the local gravitational acceleration. Thus, the point p_1 is under a horizontal tension T_1 and, likewise, p_2 is under a tension T_2 with components T_{2x} and T_{2y}. A symmetry condition around $x = x_0$ is assumed.

Figure 12.1, shows the hanging cable diagram with the aforementioned features and the involved forces.

An analysis of forces can be performed by considering T_{2x} and T_{2y} as force components, and assuming the cable is in equilibrium, we get

$$T_{2x}: \quad T_2 \cos(\theta) - T_1 = 0, \qquad T_{2y}: \quad T_2 \sin(\theta) - \lambda L g = 0. \tag{12.12}$$

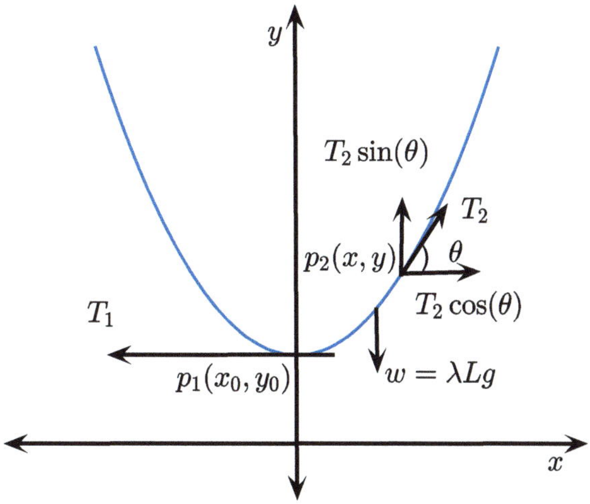

Fig. 12.1 Diagram of a hanging cable and its involved forces

From the first expression in (12.12), horizontal tension remains constant along the curve. Now, by rearranging force components and dividing one into another to dismiss variable T_2, yields,

$$\frac{\lambda g L}{T_1} = \tan(\theta), \tag{12.13}$$

where θ represents the tension angle of T_2 at an arbitrary horizontal location x. Thus, it is possible to indicate $\tan(\theta)$ in terms of a derivative or slope dy/dx, so the differential equation of the catenary curve is

$$\frac{L}{a} = \frac{dy}{dx}, \tag{12.14}$$

where $a = T_1/\lambda g$. To avoid unit inconsistency issues, variables x, y, and L are normalized by using a as the characteristic length, this is

$$\tilde{x} = \frac{x - x_0}{a}, \quad \tilde{y} = \frac{y - y_0}{a}, \quad \tilde{L} = \frac{L}{a}. \tag{12.15}$$

Therefore, the dimensionless differential equation is achieved

$$\tilde{L} = \frac{d\tilde{y}}{d\tilde{x}}, \quad \tilde{y}(0) = 0. \tag{12.16}$$

Now, by deriving both sides of (12.16) with respect to $\tilde{x}$, yields

$$\frac{d\tilde{L}}{d\tilde{x}} = \frac{d^2\tilde{y}}{d\tilde{x}^2}, \qquad \tilde{y}(0) = 0, \tag{12.17}$$

On the other hand, the dimensionless arc length $\tilde{L}$ of $\tilde{y}(\tilde{x})$ in the interval $[0, \tilde{x}]$ is given by

$$\tilde{L} = \int_0^{\tilde{x}} \sqrt{1 + \left(\frac{d\tilde{y}}{d\chi}\right)^2} d\chi, \tag{12.18}$$

which can be written in derivative form by employing the fundamental theorem of calculus. So, by considering the length formula for a curve, we have

$$\frac{d\tilde{L}}{d\tilde{x}} = \sqrt{1 + \left(\frac{d\tilde{y}}{d\tilde{x}}\right)^2}. \tag{12.19}$$

Hence, we find the second-order derivative of $\tilde{y}$ via (12.17) with (12.19), such as

$$\frac{d^2\tilde{y}}{d\tilde{x}^2} = \sqrt{1 + \left(\frac{d\tilde{y}}{d\tilde{x}}\right)^2}. \tag{12.20}$$

Equation (12.20) is non-linear and describes the cable shape along the interval $\overline{p_1 p_2}$. Hence, to achieve a function of $\tilde{y}$, the differential equation in (12.20) is rewritten in terms of u, since $u = d\tilde{y}/d\tilde{x}$. By integrating, we get

$$\sinh^{-1}(u) = \tilde{x}, \tag{12.21}$$

which can be expressed as follows,

$$\frac{d\tilde{y}}{d\tilde{x}} = \sinh(\tilde{x}), \quad \tilde{y}(0) = 0. \tag{12.22}$$

Now, by integrating once more, we have the dimensionless solution for the catenary curve, as

$$\tilde{y}(\tilde{x}) = \cosh(\tilde{x}) - 1. \tag{12.23}$$

Scaling this expression with (12.15), we get

$$y = y_0 + a\left[\cosh\left(\frac{x - x_0}{a}\right) - 1\right]. \tag{12.24}$$

12.4 Non-integer Catenary Curve

To find the fractional solution of the catenary curve, a procedure similar to that of the traditional catenary is carried out (12.14)–(12.22), but using a fractional differential operator. The dimensionless fractional differential equation for the catenary curve was given in [31],

$$\frac{d^{\alpha}\tilde{y}}{d\tilde{x}^{\alpha}} = \sinh(\tilde{x}), \quad \tilde{y}(0) = 0, \tag{12.25}$$

where $0 < \alpha \leq 1$. This expression does not depend on the definition of a specific fractional derivative, so it is possible to choose one of the fractional operators and solve it. As mentioned before, in this work we use three different types of derivatives, namely; Caputo-Dzhrbashyan and Atangana-Baleanu fractional derivatives, as well as the conformable derivative, to find different solutions of the catenary curve.

12.4.1 Fractional Catenary Based on CD Definition

Substituting the CD fractional operator (12.1) in (12.25), the expression takes the form

$$^{\text{CD}}D^{\alpha}_{\tilde{x}}\tilde{y} = \sinh(\tilde{x}), \quad \tilde{y}(0) = 0, \quad \text{and} \quad 0 < \alpha \leq 1. \tag{12.26}$$

Then, by applying the Laplace transform on both sides of the fractional differential equation (12.26) and solving for $\tilde{Y}(s)$, we obtain

$$\tilde{Y}(s) = \frac{s^{-\alpha}}{(s^2 - 1)}. \tag{12.27}$$

Employing the relationship (12.7), we find the solution as

$$\tilde{y} = \tilde{x}^{1+\alpha} E^{1}_{2,2+\alpha}\left(\tilde{x}^2\right). \tag{12.28}$$

Subsequently, this solution (12.28) is put into dimensional form using (12.15),

$$y = y_0 + a^{-\alpha}(x - x_0)^{1+\alpha} E^{1}_{2,2+\alpha}\left\{\left(\frac{x - x_0}{a}\right)^2\right\}, \quad 0 < \alpha \leq 1. \tag{12.29}$$

12.4.2 Fractional Catenary Based on ABC Definition

Likewise, by applying the ABC operator in (12.3) to the fractional problem from (12.25), we get

$$^{\text{ABC}}D^{\alpha}_{\tilde{x}}\tilde{y} = \sinh(\tilde{x}). \tag{12.30}$$

So, applying the Laplace transform, we find

$$\frac{B(\alpha)}{1-\alpha} \cdot \frac{s^{\alpha}\tilde{Y}(s)}{s^{\alpha} + \frac{\alpha}{1-\alpha}} = \frac{1}{s^2 - 1}. \tag{12.31}$$

Since, $B(\alpha) = 1$, as mentioned above. Thus, we obtain the inverse Laplace transform by using (12.7), such as

$$\tilde{y} = (1-\alpha)\sinh(\tilde{x}) + \alpha\tilde{x}^{1+\alpha} E^1_{2,2+\alpha}\left(\tilde{x}^2\right). \tag{12.32}$$

Similarly, the dimensional solution (12.32) is achieved from (12.15), as

$$\begin{aligned} y = y_0 + a\Bigg[&(1-\alpha)\sinh\left(\frac{x-x_0}{a}\right) \\ &+\alpha\left(\frac{x-x_0}{a}\right)^{1+\alpha} E^1_{2,2+\alpha}\left\{\left(\frac{x-x_0}{a}\right)^2\right\}\Bigg]. \end{aligned} \tag{12.33}$$

12.4.3 Fractional Catenary Based on Conformable Definition

Using the definition of conformal derivative (12.9) together with the expression (12.11), then (12.25) is written as

$$T_\alpha(\tilde{y})(\tilde{x}) = \tilde{x}^{\alpha-1}\frac{d\tilde{y}}{d\tilde{x}} = \sinh(\tilde{x}). \tag{12.34}$$

Subsequently, we easily find an explicit form for $\tilde{y}(\tilde{x})$, as shown

$$\tilde{y} = \int_0^{\tilde{x}} \chi^{1-\alpha}\sinh(\chi)\, d\chi. \tag{12.35}$$

Then, integrating by parts, we get

$$\tilde{y} = \tilde{x}^{1-\alpha}\cosh(\tilde{x}) - (1-\alpha)\int_0^{\tilde{x}} \chi^{-\alpha}\cosh(\chi)\, d\chi. \tag{12.36}$$

Now, we integrate again by parts the second right-hand-side term of (12.36). This process is carried out recurrently, such as

$$\tilde{y} = \tilde{x}^{1-\alpha}\cosh(\tilde{x}) - (1-\alpha)\tilde{x}^{-\alpha}\sinh(\tilde{x}) + (1-\alpha)(-\alpha)\tilde{x}^{-\alpha-1}\cosh(\tilde{x}) - \cdots, \tag{12.37}$$

which leads us to a well-defined pattern, as

$$\tilde{y} = \sum_{k=0}^{\infty}(-1)^k\frac{\Gamma(2-\alpha)}{\Gamma(2-\alpha-k)}\tilde{x}^{(1-\alpha-k)}\left.\frac{d^k}{d\chi^k}\cosh(\chi)\right|_{\chi=\tilde{x}}. \tag{12.38}$$

Recall the Gamma function property $n! = \Gamma(n+1)$ and the periodic nature of hyperbolic functions. So, we can reduce even and odd terms of summation in (12.37), as follows

$$(-1)^k \frac{d^k}{d\chi^k} \cosh(\chi)\bigg|_{\chi=\tilde{x}} = \begin{cases} \cosh(\tilde{x}), & \text{if } k = 0, 2, 4, \ldots, \\ -\sinh(\tilde{x}), & \text{otherwise.} \end{cases} \tag{12.39}$$

Therefore, Eq. (12.37) is rewritten in two parts, as given

$$\begin{aligned} \tilde{y}(\tilde{x}) = \cosh(\tilde{x}) \sum_{k=0,2,\cdots}^{\infty} \frac{\Gamma(2-\alpha)}{\Gamma(2-\alpha-k)} \tilde{x}^{(1-\alpha-k)} - \\ - \sinh(\tilde{x}) \sum_{k=1,3,\cdots}^{\infty} \frac{\Gamma(2-\alpha)}{\Gamma(2-\alpha-k)} \tilde{x}^{(1-\alpha-k)}. \end{aligned} \tag{12.40}$$

By changing summation indices to follow the same sequence, we have

$$\tilde{y} = \tilde{x}^{1-\alpha} \cosh(\tilde{x}) \sum_{n=0}^{\infty} \frac{\alpha(2-\alpha)\tilde{x}^{-2n}}{\Gamma(2-\alpha-2n)} - \tilde{x}^{-\alpha} \sinh(\tilde{x}) \sum_{n=0}^{\infty} \frac{\Gamma(2-\alpha)\tilde{x}^{-2n}}{\Gamma(1-\alpha-2n)}. \tag{12.41}$$

It should be noted that these infinite series are directly related to the ML function on (12.6), so we can write the dimensionless conformable catenary equation as

$$\tilde{y} = \Gamma(2-\alpha) \left[\tilde{x}^{2-\alpha} \cosh(\tilde{x}) E_{-2,2-\alpha}(\tilde{x}^{-2}) - \tilde{x}^{-\alpha} \sinh(\tilde{x}) E_{-2,1-\alpha}(\tilde{x}^{-2}) \right]. \tag{12.42}$$

Finally, by employing the property for ML functions in (12.8) and rescaling $\tilde{y}(\tilde{x})$ to $y(x)$, we obtain the catenary curve according to the Conformable derivative

$$\begin{aligned} y(x;\alpha) = y_0 + a\Gamma(2-\alpha) \Bigg[\left(\frac{x-x_0}{a}\right)^{(2-\alpha)} E_{2,3-\alpha}\left(\left(\frac{x-x_0}{a}\right)^2\right) \sinh\left(\frac{x-x_0}{a}\right) \\ - \left(\frac{x-x_0}{a}\right)^{(3-\alpha)} E_{2,4-\alpha}\left(\left(\frac{x-x_0}{a}\right)^2\right) \cosh\left(\frac{x-x_0}{a}\right) \Bigg]. \end{aligned} \tag{12.43}$$

12.5 Results

In the present work, we have analized three models based on the catenary curve and fractional derivatives: the Caputo fractional definition—*i.e.*, CD, the Caputo Fabrizio derivative (CF) from the previous work [31], and ABC—, and another based on the Conformable derivative. We employed parameters values such as $h = 5$ m, $x_0 = 0$ m, $a = T_1/\lambda g = 10$ m, and $x \in [-20, 20]$ m. Also, we varied γ between 0.7 and 1.00, where this upper bound corresponds to the ordinary differential order related to the traditional catenary curve. Then, the proposed models were plotted in normalized scale to facilitate visual comparison.

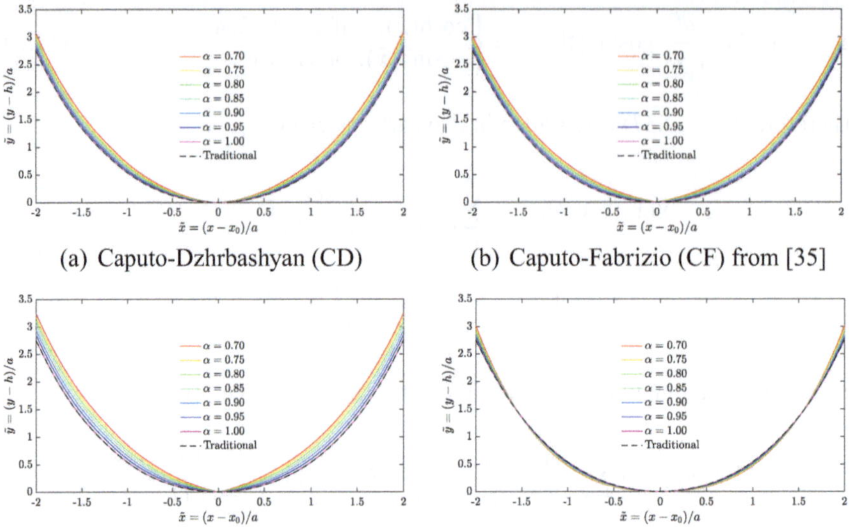

(a) Caputo-Dzhrbashyan (CD)

(b) Caputo-Fabrizio (CF) from [35]

(c) Atangana-Baleanu-Caputo (ABC)

(d) Conformable

Fig. 12.2 Catenary curves using different operators and varying the order α

in adittion, the result based on the work [31], which used the CF was

$$y(x) = y_0 + a\left[(1-\alpha)\sinh\left(\frac{x-x_0}{a}\right) + \alpha\cosh\left(\frac{x-x_0}{a}\right) - \alpha\right] \tag{12.44}$$

Figure 12.2, presents the catenary curves determined through the aforementioned fractional operators. First of all, we remark that for all the models when $\alpha \to 1$ their solutions tend to the ordinary solution, as expected. Plots show that all the Caputo-based fractional curves are over the traditional catenary curve. This pattern is also observed with the Confomable-based trajectories when $|\tilde{x}| \gtrapprox 1.5$, but those are under the traditional catenary when $|\tilde{x}| \lessapprox 1.5$. Then, the curves achieved with $\alpha = 1$ are drawn as black dashed lines.

12.6 Conclusions

Using the CD and ABC definitions of fractional derivatives, as well as the conformable derivative, a complete theoretical analysis of the catenary has been obtained. The result presented by [32] converge to the solution using CD fractional derivative using the properties of the ML function. Various numerical simulations were performed varying the fractional order α. The models obtained here were also compared with the one obtained in [31]. The resulting curves showed that all fractional catenaries tend to the traditional catenary as $\alpha \to 1$, and these catenaries differ in at least two ways as α decreases. The first was observed in Caputo-based fractional catenaries (CD, CF and ABC) that exhibit a difference trend similar to the traditional catenary exponential as $|\tilde{x}|$ increases and α decreases. This pattern is smooth on curves based on the CD definition and sharp and scaled on those based on CF and AB. Furthermore, the conformable model showed closed curves around the traditional catenary with a bounded difference of $0.0 \leq \Delta\tilde{y} < 0.1$ when $|\tilde{x}| \lessapprox 1.5$.

Acknowledgements The authors thank Department of Multidisciplinary Studies, Departament of Electrical Engineering of DICIS University of Guanajuato, Escuela de Ingeniería y Ciencias of Tecnológico de Monterrey and Departament of Mechatronics, TecNM/Irapuato for their support.

References

1. Li Z, Lee TU, Pietroni N, Snooks R, Xie YM (2024) Design and construction of catenary-ruled surfaces. Structures 59(105755):105755. https://doi.org/10.1016/j.istruc.2023.105755
2. Goldstein G, Kapadia Y, Lin TY, Zhivago P (2015) The catenary curve: a guide to mandibular arch form. Periodontics Prosthodontics 01(01). https://doi.org/10.21767/2471-3082.100001
3. Aljayousi M, Al-Khateeb S, Badran S, Alhaija ESA (2021) Maxillary and mandibular dental arch forms in a Jordanian population with normal occlusion. BMC Oral Health 21(1). https://doi.org/10.1186/s12903-021-01461-y
4. Luo X (2019) Catenary optics, 1st edn. Springer
5. Kim J, An D (2009) Evaluation of progressive collapse potential of steel moment frames considering catenary action. Struct Des Tall Spec Build 18(4):455–465. https://doi.org/10.1002/tal.448
6. Monje CA, Chen YQ, Vinagre BM, Xue D, Feliu V (2010) Fractional-order systems and controls. advances in industrial control. Springer, Berlin, Series
7. Caponetto R, Dongola G, Fortuna L, Petrás I (2010) Fractional order systems: modelling and control applications. World Scientific
8. Golmankhaneh AK, Lambert L (2012) Investigations in dynamics: with focus on fractional dynamics. Academic Publishing
9. Baleanu D, Günvenc ZB, Tenreiro Machado JA (eds) (2010) New trends in nanotechnology and fractional calculus applications. Springer
10. Wyss W (1986) Fractional diffusion equation. J Math Phys 27:2782–2785
11. Westerlund S (1994) Capacitor theory. IEEE Tran Die Elec Ins 1(10):826–839
12. Hermann R (2011) Fractional calculus. World Scientific
13. Murcia-Sepúlveda N, Cruz-Duarte JM, Martin-Diaz I, García-Pérez A, Rosales-García JJ, Avina-Cervantes JG, Correa-Cely CR (2019) Fractional calculus-based processing for feature extraction in harmonic polluted fault monitoring systems. Energies 12(19):3736. https://doi.org/10.3390/en12193736
14. Martínez-Jiménez L, Cruz-Duarte JM, Rosales-García JJ, Cruz-Aceves I (2018) Enhancement of vessels in coronary angiograms using a Hessian matrix based on Grunwald-Letnikov fractional derivative. In: Proceedings of the 8th international conference on biomedical engineering and technology. ACM, New York, pp 51–54
15. Cruz-Duarte JM, Rosales-García JJ, Correa-Cely CR, García-Perez A, Avina-Cervantes JG (2018) A closed form expression for the Gaussian-based Caputo-Fabrizio fractional derivative for signal processing applications. Commun Nonlinear Sci Numer Simulat 61:138–148

16. Aguirre-Ramos H, Avina-Cervantes JG, Cruz-Aceves I, Ruiz-Pinales J, Ledesma S (2018) Blood vessel segmentation in retinal fundus images using Gabor filters, fractional derivatives, and expectation maximization. App Math Comp 339:568–587. https://doi.org/10.1016/j.amc.2018.07.057
17. Lozynsky A, Chaban A, Perzynski T, Szafraniec A, Kasha L (2021) Application of fractional-order calculus to improve the mathematical model of a two–mass system with a long shaft. Energy 14(1854). https://doi.org/10.3390/en14071854
18. Coronel-Escamilla A, Lavín-Delgado JE, Gómez-Aguilar JF, Torres L (2020) Fractional dynamics and synchronization of Kuramoto oscillators with nonlocal, nonsingular and strong memory. Alex Eng J 59(4):1941–1952. https://doi.org/10.1016/j.aej.2019.12.015
19. García-Aspeitia MA, Fernandez-Anaya G, Hernández-Almada A, Leon G, Magaña J (2022) Cosmology under the fractional calculus approach. Mont Not Ro Astr Soc Stac 3006. https://doi.org/10.1093/mnras/stac3006
20. Li X, Wang Y (2022) Numerical simulation of the fractional–order Rössler chaotic systems with Grünwald Letnikov fractional derivative. Fractals 30(8),2240229
21. Wu X, Huang Y (2022) Adaptive fractional-order non-singular terminal sliding mode control based on fuzzy wavelet neural networks for omnidirectional mobile robot manipulator. ISA Trans 121:258–267. https://doi.org/10.1016/j.isatra.2021.03.035
22. Piotrowska E, Rogowski K (2021) Time-domain analysis of fractional electrical circuit containing two ladder elements. Electr 10(475). https://doi.org/10.3390/electronics10040475
23. Capelas de Oliveira E, Tenreiro Machado JA (2014) A review of definitions for fractional derivatives and integrals. Math Probl Eng ID:238459
24. Oldham KB, Spanier J (eds) (1974) Chapter 1: introduction. In: Mathematics in science and engineering, vol 111. Academic press, pp 1–24
25. Samko SG, Kilbas AA, Marichev OI (1993) Fractional integrals and derivatives: theory and applications. Gordon and Breach
26. Podlubny I (1999) Fractional differential equations. Academic Press
27. Uchaikin V (2013) Fractional derivatives for physicists and engineers. Springer
28. Caputo M, Fabrizio M (2015) A new definition of fractional derivative without singular kernel. Progr Fract Differ Appl 1:73–85
29. Atangana A, Baleanu D (2016) New fractional derivatives with non-local and non-singular kernel, theory and application to heat transfer model. Therm Sci 20(2):763
30. Khalil R, Horani M, Yousef A, Sababheh M (2014) A new definition of fractional derivative. J Comp Appl Math 264:65–70
31. Martínez-Jiménez L, Cruz-Duarte JM, Rosales-García JJ (2019) Fractional solution of the catenary curve. Math Methods Appl Sci 44(10):7969–7978. https://doi.org/10.1002/mma.5608
32. Becerra-Guzmán G, Villa-Morales J (2024) Erroneous applications of fractional calculus: the catenary as a prototype. Mathematics 12(14):2148. https://doi.org/10.3390/math12142148
33. Kochubei AN (2011) General fractional calculus evolution equations, and renewal processes. Integr Eqn Oper Theory 71(4):583–600
34. Mittag-Leffler GM (1903) Sur la nouvelle fonction $E_\alpha(x)$. Comptes Rendus de l' Academie des Sciences Paris 137:554–558
35. Mittag-Leffler GM (1903) Une generalisation de l' integrale de Laplace-Abel. Comptes Rendus de l' Academie des Sciences Paris Serie II(137):537–539
36. Mittag-Leffler GM (1905) Sur la representation analytiqie d'une fonction monogene (cinquieme note). Acta Math 29(1):101–181
37. Xiangang L (2019) Catenary optics. Springer Nature
38. Kim J, An D (2009) Evaluation of progressive collapse potential of steel moment frames considering catenary action. Struct. Des Tall Spec Build 18:455–465. https://doi.org/10.1002/tal.448
39. Chatterjee N, Nita BG (2010) The hanging cable problem for practical applications. Atlantic Electron J Math 4(1):70–77

40. Andreu A, Gil L, Roca P (2006) A new deformable catenary element for the analysis of cable net structure. Comput Struct 84:1882–1890
41. Goldstein G, Kapadia YY, Lin TY, Zhivago P (2015) The catenary curve: a guide to mandibular arch form. iMedPub J 1:1–4. https://doi.org/10.21767/2471-3082.100001
42. MacConaill MA, Scher EA (1949) The ideal form of human dental arcade, with some prosthetic applications. Dent Record 69(11):285–302
43. Scott JH (1957) The shape of the dental arches. J Dent Res 36:996–1003
44. Currier JH (1969) A computational geometric analysis of human dental arch form. Am J Orthod 56:164–179
45. Lee J, Chen S, Reece GP, Crosby MA, Beahm EK, Markey MK (2012) A novel quantitative measure of breast curvature based on catenary. IEEE Trans Biomed Eng 59(4):1115–1124
46. Mingbo P, Xiong L, Xiaoliang M, Yanquin W, Zeyu Z et al (2015) Catenary optics for achromatic generation of perfect optical angular momentum. Sc Adv 1(9). https://doi.org/10.1126/sciadv.1500396

Chapter 13
Modeling a Real Electrical Circuit with Fractional Calculus

J. Juan Rosales, Sergio Israel Velázquez Chávez, and F. A. Godínez

13.1 Introduction

Fractional calculus (FC), which might be considered a generalization of ordinary calculus deals with derivatives and integrals of non-integer order [1, 2]. Although fractional derivatives and integrals appeared almost at the same time as ordinary derivatives, it was not until about 40 years ago that they gained wide acceptance in the scientific community, particularly in applied sciences. It has been shown that the solutions obtained from fractional differential equations fit better the experimental data [3–9]. However, despite the efforts to have a physical and geometrical interpretation of fractional derivatives and integrals, today there is no fully accepted answer to this issue. A further aspect that requires attention and that is overlooked in many works is the dimensional homogeneity of the equations, when particularly physical processes are modeled. It is very common to find in the literature equations that do not comply with the principle of homogeneity [10–12], which states that dimensions of each of the terms of a dimensional equation on both sides should be the same [13, 14]. The standard procedure for converting an integer order model into its corresponding

J. Juan Rosales (✉)
División de Ingenierías Campus Irapuato-Salamanca, Carretera Salamanca-Valle de Santiago, Salamanca, Guanajuato, Mexico
e-mail: rosales@ugto.mx

S. I. Velázquez Chávez
Instituto Tecnológico de la Laguna del Tecnológico Nacional de Mexico, Instituto Tecnológico de la Laguna s/n, Primero de Cobián, Centro, Torreón, Coahuila, Mexico
e-mail: d.sivelazquezc@lalaguna.tecnm.mx

F. A. Godínez
Instituto de Ingeniería, Universidad Nacional Autónoma de México, Unidad de Investigación y Tecnología Aplicadas, Apodaca, Nuevo León, Mexico
e-mail: fgodinezr@iingen.unam.mx

J. J. Rosales García et al. (eds.), *Advances in Fractional Calculus*,
Modeling and Optimization in Science and Technologies 21,
https://doi.org/10.1007/978-3-031-83989-4_13

fractional one, is to replace the classical differential operator(s) with those of non-integer order. This method, however, unbalances the physical units (time, length, mass, etc.) in the fractional equations. One possibility to overcome this difficulty is to introduce into the unbalanced equation(s) a parameter, commonly denoted as σ, firstly introduced in [15, 16]. The units of this parameter depend on the variable to which the derivative is applied (time or space). Thus, if we are dealing with a dynamic process where time derivative is required, σ must have units of time. The above strategy has yielded dimensionally consistent results in models of different processes in the fields of electric circuits, physics, biology and engineering, among others [17–25]. This article develops the fractional version of the integer-order model that describes the charging and discharging process of an RC circuit. The solution of the fractional model, based on the Caputo derivative, is compared with the solution of the classical integer-order model and with experimental data. When the circuit components are new, there is good agreement between the experimental data and the integer-order model. However, when the capacitor has undergone multiple charging and discharging cycles over an extended period, a phenomenon of aging occurs in the circuit (both in the resistor and in the capacitor). In this case, a slight discrepancy is observed between the actual behavior of the circuit during charging and the predictions of the classical model. In contrast, when employing the fractional model with a derivative order close to 1, a better alignment between experimental results and the model is achieved. The fractional model can capture the aging of the circuit, which is expected due to the ability of the non-local Caputo operator to account for memory effects in dynamic processes.

13.2 Preliminary Concepts

There are some definitions of fractional derivatives, the most popular are; Riemann-Liouville, Caputo, Caputo-Fabrizio, Atangana-Baleanu and conformable. In this work we will use the Caputo fractional derivative, defined as [1],

$$ {}^{C}D_t^{\gamma} f(t) = \frac{1}{\Gamma(1-\gamma)} \int_0^t \frac{f'(\tau)}{(t-\tau)^{\gamma}} d\tau, \qquad 0 < \gamma \leq 1, \tag{13.1} $$

and its corresponding Laplace transform is given by

$$ \mathcal{L}\left[{}^{C}D^{\gamma} f(t)\right] = s^{\gamma} F(s) - s^{\gamma-1} f(0). \tag{13.2} $$

The corresponding generalization of the exponential function is the Mittag-Leffler funtion defined as

$$ \mathbb{E}_{\gamma}(-at^{\gamma}) = \sum_{n=0}^{\infty} \frac{(-at^{\gamma})^n}{\Gamma(\gamma n + 1)}. \tag{13.3} $$

In the particular case $\gamma = 1$, the expression (13.3) reduces to the exponential function

$$\mathbb{E}_\gamma(-at^\gamma)\Big|_{\gamma=1} = \sum_{n=0}^{\infty} \frac{(-at)^n}{\Gamma(n+1)} = e^{-at}. \tag{13.4}$$

The Mittag-Leffler function plays a crucial role in the solution of linear fractional differential equations, just as the exponential function plays in linear ordinary differential equations with constant coefficients.

The Laplace transform of the Mittag-Leffler function is given by

$$\int_0^\infty e^{-st} t^{\gamma m+\beta-1} \mathbb{E}^{(m)}_{\gamma,\beta}(\pm at^\gamma)dt = \frac{m! s^{\gamma-\beta}}{(s^\gamma \mp a)^{m+1}}. \tag{13.5}$$

Consequently, its inverse is

$$\mathcal{L}^{-1}\left[\frac{m! s^{\gamma-\beta}}{(s^\gamma \mp a)^{m+1}}\right] = t^{\gamma m+\beta-1} \mathbb{E}^{(m)}_{\gamma,\beta}(\pm at^\gamma). \tag{13.6}$$

This expression is very useful for obtaining analytical solutions.

13.2.1 Ordinary RC Circuit

We start with the RC electrical circuit, Fig. 13.1.

The differential equation describing an unloaded RC circuit with constant $E(t) = V_s$ and initial condition $V(0) = 0$, is

$$\frac{dV}{dt} + \frac{1}{\tau} V(t) = \frac{V_s}{\tau}, \qquad \tau = RC. \tag{13.7}$$

The solution of the equation (13.7) is well known and it has the form

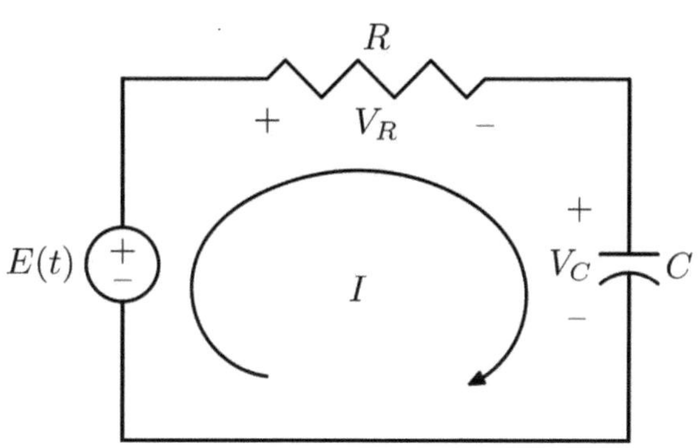

Fig. 13.1 RC circuit

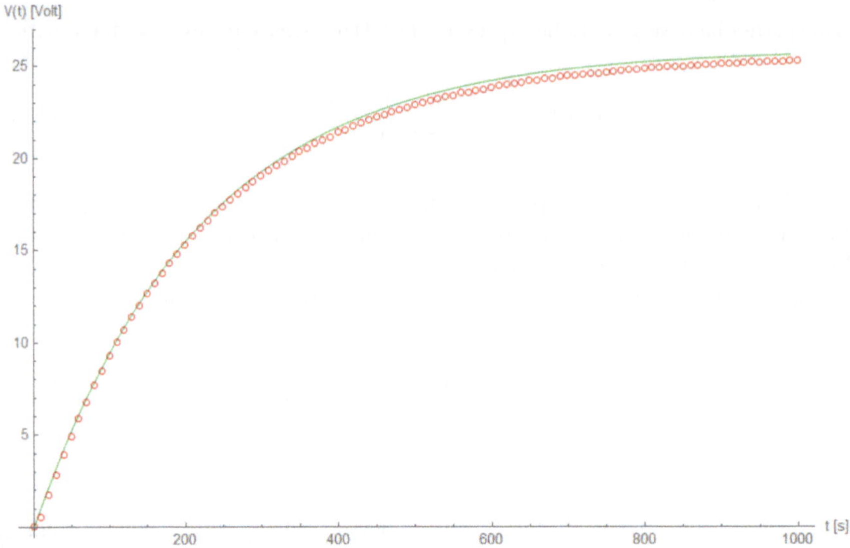

Fig. 13.2 Comparison between experimental data (hollow symbols) and the classical solution (13.8) of the integer order solution (solid line)

$$V(t) = V_s\left(1 - e^{-\frac{t}{\tau}}\right). \tag{13.8}$$

We can see that the ordinary differential equation (13.7) is dimensionally homogeneous, since on the right and left of the Eq. (13.7), we have $volts/s$. As expected, its corresponding solution (13.8) is also dimensionally consistent, since on the right and left of the Eq. (13.8), we have $volts$. The above is very important in applied sciences and engineering [13, 14].

The Eq. (13.7) describes the behavior (charge process in our case) of an RC circuit with a constant source, where its components are ideal. The solution (13.8) exhibits an asymptotic exponential decay, when $t \to \infty$ the voltage $V(t \to \infty) \to V_s$, a constant.

13.2.1.1 Fractional RC Circuit

We transform the ordinary differential equation (13.7) into a fractional one. We follow the systematic way given in [15, 16], then we have

$$\frac{d}{dt} = \frac{1}{\sigma^{1-\gamma}} \frac{d^\gamma}{dt^\gamma}, \qquad 0 < \gamma \leq 1, \tag{13.9}$$

where σ, in this case, has dimension of seconds. Then, the corresponding fractional differential equation has the form

$$\frac{d^{\gamma} V}{dt^{\gamma}} + aV(t) = b, \tag{13.10}$$

where $a = \frac{\sigma^{1-\gamma}}{\tau}$ and $b = \frac{\sigma^{1-\gamma}}{\tau} V_s$. Equation (13.10) is dimensionally homogeneous, it has been obtained and studied for different types of fractional derivatives, such as; Caputo, Caputo-Fabrizio, Atangana-Baleanu fractional derivatives and the conformal derivative [18], where σ was restricted to take the value of RC. It was observed that, if one applies the Caputo-Fabrizio and Atangana-Baleanu fractional derivatives to Eq. (13.10) and solves it, then, dimensionally inhomogeneous solutions are obtained. From a physical point of view, this is a serious problem, since in the solution we would have, on one hand, voltage and on the right something different. A similar work has been published [9], where they have used the Planck time $t_p = \sqrt{\frac{\hbar G}{c^5}}$ instead of σ in (13.9). Planck time is a constant and we do not see the physical reason for using it.

Then, applying the Caputo fractional derivative, (13.1), and the Laplace transform, (13.2), with the initial condition $V(0) = 0$, in (13.10), we have

$$s^{\gamma} V(s) + aV(s) = \frac{b}{s}. \tag{13.11}$$

Solving with respect to $V(s)$, we get,

$$V(s) = \frac{b}{s(s^{\gamma} + a)}. \tag{13.12}$$

The inverse Laplace transform of (13.12), gives the solution

$$V(t;\gamma) = V_s\Big[1 - \mathbb{E}_{\gamma}\Big(-\frac{\sigma^{1-\gamma}}{\tau} t^{\gamma}\Big)\Big]. \tag{13.13}$$

As we can see, this solution is dimensionally consistent, since the argument of the Mittag-Leffler function is dimensionless.

In [23], the parameter γ, which represents the order of fractional differential equation was related to the parameter σ, as follows

$$\gamma = \frac{\sigma}{RC} \quad \Longrightarrow \quad 0 < \sigma \leq RC. \tag{13.14}$$

Then, the Eq. (13.13) reduces to

$$V(t;\gamma) = V_s\Big[1 - \mathbb{E}_{\gamma}\Big[-\gamma^{1-\gamma}\Big(\frac{t}{\tau}\Big)^{\gamma}\Big]\Big]. \tag{13.15}$$

Aslo, we can define a new dimensionless magnitude

$$\delta = 1 - \gamma, \tag{13.16}$$

that characterizes the presence of fractional components in the system. With this in mind, the expression (13.13), takes the form

$$V(t;\delta) = V_s \left\{ 1 - \mathbb{E}_{1-\delta}\left[-(1-\delta)^{\delta}\left(\frac{t}{\tau}\right)^{1-\delta} \right] \right\}, \qquad \delta < 1. \tag{13.17}$$

In the particular case $\gamma = 1$, we have from de expressions (13.13) (13.15) and (13.17), the ordinary solution (13.8),

$$V(t) = V_s\left[1 - \mathbb{E}_1\left(-\frac{t}{\tau}\right)\right] = V_s\left(1 - e^{-\frac{t}{\tau}}\right), \qquad \delta = 0. \tag{13.18}$$

From $\delta = 0$, follows that $\gamma = 1$ and $\sigma = RC$. In general, $0 \leq \delta < 1$.

13.2.1.2 Experiment

In this section, the measurement of wear during the charging and discharging of a capacitor is described through experimental analysis to determine how the characteristics of the circuit and the operating time affect the capacitance and efficiency of the capacitor.

A. **Description**: Using a data acquisition card, the continuous charging and discharging of a capacitor is carried out, with a period of 3 s for each action. These periods are controlled by activating relays through the same card. The data was recorded six times a week every 24 h and stored through programming in an Excel file.

B. **Materials**: The chosen capacitor is 2200 μF. For the charging and discharging circuits, 100 Ω resistors are used. The power supply is provided by a variable bipolar source. The relays for the selection of charging and discharging are 5 V and 10 A. Additionally, there is a 12 kΩ resistor and another one of 49 kΩ.

13.2.1.3 Design of the Circuit and Charging Discharging program

The capacitor is connected to a 'CHARGE' circuit and a 'DISCHARGE' circuit. The 'CHARGE' circuit consists of a relay that has a power supply connected to its 'COMMON' terminal, which provides 25.9 V for the operation of the circuit. Meanwhile, the 'NORMALLY OPEN' terminal is connected to a 100 Ω resistor, which in turn connects to the positive terminal of the capacitor, while the negative terminal is connected to ground. For the 'DISCHARGE' circuit, another relay connects the positive terminal of the capacitor to its 'COMMON' terminal, and the 'NORMALLY OPEN' terminal is connected to ground. Both phases are activated by the coils of each relay, which are controlled by the data acquisition board, in this case, an ARDUINO MEGA board. It alternates the activation using its digital pins $D8$ and $D9$, respectively for 'CHARGE' and 'DISCHARGE'. This alternation between phases is carried out using a program created in the ARDUINO IDE, where the code alternates the activation of the respective charge and discharge relays and, within each activation, performs readings from pin A0 to which the positive terminal of the capacitor is connected. These readings are printed to the serial monitor of the board for later storage.

13.2.2 Data Reading

Due to the fact that data reading is performed with the same ARDUINO MEGA board, a voltage divider circuit is implemented because the board has a 10-bit ADC converter that only supports 5 V; therefore, the 25 V used would damage it.

13.2.3 Data Storage and Processing

A total of 60,000 data points from charging and discharging cycles are stored in an Excel file. To process these data, two columns are added: one for the measured voltage, calibrated based on the ADC readings from the ARDUINO MEGA, and another for the elapsed time, dividing the cycles into 3 s segments and taking readings every 10 ms. The relationship to calculate the voltage is: voltage $= value * 25.9/1023$. A scatter plot is then created to represent one charging cycle and one discharging cycle over 6 s. In Fig. 13.2, a photo of the actual circuit used in the experiments is shown.

13.3 Results and Discussion

The solutions of the fractional models, Eqs. (13.13) (13.15) and (13.17), obtained in the previous sections, were fitted to a data set found experimentally. The fitting process was performed using the least square method; accordingly, it is required to compute the error between the experimental data $(y_i)_{i=1...m}$ and the values from a model $(M_i)_{i=1...m}$ [5]. Table 13.1 summarizes the results regarding the fit errors for the three models considered. Clearly, the model corresponding to equation (13.13) is the one that best fits the experimental data for an aged capacitor with a charge and discharge period of 730.5 h. Consequently, the fractional model (Eq. (13.13)) fits better, as seen in Fig. 13.3, compared to the classical model shown in Fig. 13.4. The presented results indicate a slight discrepancy between the experimental curves

Fig. 13.3 Real RC circuit

Table 13.1 Fitting parameter values for different models

Model	γ	σ (s)	δ	Error
Classical	1	–	–	12.6529
Fractional Eq. 13.13	0.981298	48.7985	–	2.00491
Fractional Eq. 13.15	0.973492	–	–	4.54092
Fractional Eq. 13.17	–	–	0.0276738	3.95797

and the classical model. However, this discrepancy can be corrected by employing the fractional model, considering the order of the Caputo derivative to be close to 1, as shown in the Table 13.1. This finding has been reported previously [26]; however, unlike those authors, who interpret the slight discrepancy as a regenerated memory effect, we associate it with an aging process.

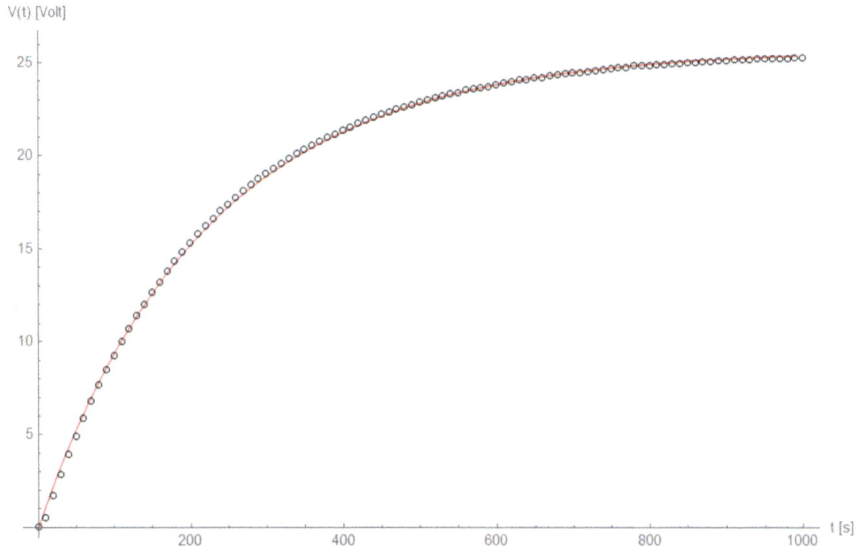

Fig. 13.4 Comparison between experimental data (hollow symbols) and the solution corresponding to Eq. (13.13), solid line

13.4 Conclusion

A theoretical and experimental study of the charging process in an RC circuit has been conducted. The classical integer-order model shows slight discrepancies with the experimental data after numerous cycles of charging and discharging over time. In contrast, the fractional model based on the Caputo derivative fits the experimental curves significantly better following the aging process of the capacitor induced by the sequence of charge/discharge cycles. This proposed fractional model is dimensionally homogeneous, which is often not the case in many works presented in the literature. A total of three variants of dimensionally consistent solutions were proposed. However, the variant incorporating the σ parameter with units of time (13.9), without any restriction [8, 27], emerges as the best option since it aligns more closely with the experimental data. We suggest that the order of the derivative may be associated with an aging process of the system, specifically linked to the non-local operator represented by the Caputo derivative. From a physical perspective, this aging can be associated with an increase in entropy; however, this requires a more in-depth investigation of the electrical system studied here and other systems of various natures, which will be explored in future research.

Acknowledgements J. Juan Rosales appreciates the support given by DAIP within the framework of the project: Sobre la dimensionalidad de los operadores diferenciales fraccionarios, Grant 203/2024.

References

1. Podlubny I (1999) Fractional differential equations. Academic Press, New York
2. Samko SG, Kilbas AA, Marichev OI (1993) Fractional integrals and derivatives: theory and applications. Gordon and Breach, New York
3. Martínez L, Jiménez JM, Cruz Duarte JJ, Escalante Martínez JJ, García R (2021) Analytical and experimental study for mechanical vibrations of a two couples spring masses system via Caputo-based derivatives. Math Meth Appl Sci. https://doi.org/10.1002/mma.7421
4. Bhangale Nikita, Kachhia Krunal B (2022) A fractional calculus approach to study Newton's law of cooling. Progr Fract Differ Appl 8(2):275–287
5. Rosales JJ, Godínez FA, Martínes L, Chávez O (2021) Combining fractional derivatives to predict blood alcohol concentration. Fractals 29(3):2150063. https://doi.org/10.1142/S0218348X21500638
6. Rosales García J, Andrade Lucio JJ, Shulika O (2020) Conformable derivative applied to experimental Newton's law of cooling. Rev Mex Fís 66(2):224–227 https://doi.org/10.31349/RevMexFis.66.224
7. lin D, Liao X, Dong L, Yang R, Yu SS, Iu HH, Fernando T, Li Z (2021) Experimental study of fractional-order RC circuit model using the Caputo and Caputo-Fabrizio derivatives. IEEE Trans Circ Syst I 68(3):1034–1044
8. Ebaid A, El-Zahar ER, Aljohani AF, Salah B, Krid M, Tenreiro Machado J (2019) Analysis of the two-dimensional fractional projectile motion in view of the experimental data. Nonlinear Dyn 97:1711–1720
9. Ertik H, Calik AE, Sirin H, Sen M, Oder B (2015) Investigation of electrical RC circuit within the framework of fractional calculus. Rev Mex Fís 61:58–63
10. Alsaedi A, Nieto JJ, Venktesh V (2015) Fractional electrical circuits. Adv Mech Eng 7(12):1–7
11. Ebaid Abdelhalim (2011) Analysis of projectile motion in view of fractional calculus. Appl Math Modell 35:1231–1239
12. Shahid A, Kamal S, Shah J, Thabet A (2023) An efficient method for the fractional electric circuits based on Fibonacci wavelet. Results Phys 52:106753
13. Dym CL (2004) Principle of mathematical modeling, 2nd edn. Elsevier Academic Press
14. Gibbings JC (2011) Dimensional analysis. Springer-Verlag, London
15. Rosales JJ, Gómez GF, Guía M, Tkach VI (2011) Fractional electromagnetic waves. In: 2011 11th international conference on laser and fiber-optical networks modeling (LFNM). https://doi.org/10.1109/LFNM.2011.6144969
16. Gómez-Aguilar JF, Rosales-García JJ, Bernal-Alvarado JJ, Córdova-Fraga T, Guzmán-Cabrera R (2012) Fractional mechanical oscillators. Rev Mex de Fís 58:348–352
17. Godinez FA, Rosales JJ, Esparza-Posadas MF (2022) Newton's law of cooling with Caputo derivative: consistent dimensionality to compare with experiments. Fractals 30:2250187
18. Rosales JJ, Filoteo JD, Gonzalez A (2018) A comparative analysis of the RC circuit with local and non-local fractional derivatives. Rev Mex de Fís 64:647–654
19. Carreño CA, Rosales JJ, Merchan LR, Lozano JM, Godínez FA (2019) Comparative analysis to determine the accuracy of fractional derivatives in modeling supercapacitors. Int J Circ Theor Appl 1–12
20. Juan Rosales J, Guía M, Martínez J, Baleanu D (2013) Motion of a particle in a resisting medium using fractional calculus approach. In: Proceedings of the Romanian Academy, Series A, vol 14, pp 42–47
21. Alharbi FM (2020) Analysis of logistic model with constant harvesting in a view of non-integer derivative. J Math Syst Sci 10. https://doi.org/10.17265/2159-5291/2020.02.004
22. Chunga WS, Zareb S, Hassanabadi H, Kriz J, Maghsoodid E (2020) The investigation of a classical particle in the presence of fractional calculus. Rev Mex Fís 66(6):840–847
23. Gómez F, Rosales J, Guía M (2013) RLC electrical circuit of non-integer order. Cent Eur J Phys. https://doi.org/10.2478/s11534-013-0265-6

24. Baleanu D, Hassan Abadi M, Jajarmi A, Zarghami Vahid K, Nieto JJ (2022) A new comparative study on the general fractional model of COVID-19 with isolation and quarantine effects. Alex Eng J 61:4779–4791
25. Gómez F, Bernal J, Rosales J, Córdova T (2012) Modeling and simulation of equivalent circuits in description of biological systems: a fractional calculus approach. J Electr Bioimp 3:2–11
26. Uchaikin V, Sibatov R, Uchaikin D (2009) Memory regeneration phenomenon in dielectrics: the fractional derivative approach. Physica Scripta 2009:014002
27. Ebaid A, Masaedeh B, El-Zahar E (2017) A new fractional model for the falling body problem. Chin Phys Lett 34(2):020201

24. Baleanu D, Hassan Abadi M, Jajarmi A, Zarghami Vahid K, Nieto JJ (2022) A new comparative study on the general fractional model of COVID-19 with isolation and quarantine effects. Alex Eng J 61:4779–4791
25. Gómez F, Bernal J, Rosales J, Córdova T (2012) Modeling and simulation of equivalent circuits in description of biological systems - a fractional calculus approach. J Electr Bioimp 3:2–11
26. Uchaikin V, Ambrozov V, [illegible] D (2009) Memory regeneration phenomenon in dielectrics: fractional derivative approach. Physica Scripta 2009:T136
27. [illegible] (2017) A new fractional model for the falling body problem. Chin Phys Lett 34:030201

MIX
Papier aus verantwortungsvollen Quellen
Paper from responsible sources
FSC® C105338

If you have any concerns about our products,
you can contact us on
ProductSafety@springernature.com

In case Publisher is established outside the EU,
the EU authorized representative is:
Springer Nature Customer Service Center GmbH
Europaplatz 3, 69115 Heidelberg, Germany

Printed by Libri Plureos GmbH
in Hamburg, Germany